I0030111

Data Analytics and Machine Learning for Students

Using Python

Mike Gold

Data Analytics and Machine Learning for Students

Using Python

Mike Gold

ISBN 979-8-9915657-3-8

Leanpub

This is a Leanpub book. Leanpub empowers authors and publishers with the Lean Publishing process. Lean Publishing is the act of publishing an in-progress ebook using lightweight tools and many iterations to get reader feedback, pivot until you have the right book and build traction once you do.

© 2025 Mike Gold

Also By Mike Gold

Veil of Blood and Magic
C# Evolution
Epic Python Coding: Interactive Coding Adventures for Kids
One of Me
Crafting Applications with ChatGPT API
Creating Video Games Using PyGame
Creating a Wordle Game in React and TypeScript

Contents

CONTENTS

Who Is This Book For?

This book is designed for high school students (Grades 9–12) who are curious about the exciting fields of data analytics and machine learning. It is ideal for students who:

- Are new to the concepts of data analytics and machine learning and want to explore how data shapes our world.
- Are taking introductory courses in computer science, statistics, or math and want to supplement their learning with practical projects.
- Have some experience with coding or data analysis and want to deepen their understanding of how data is used to make informed decisions.

No prior knowledge of data science or machine learning is required, but familiarity with basic algebra and an interest in working with data will help students get the most out of this book.

In addition to students, this book is also a valuable resource for:

- **Educators**: Teachers can use this book to introduce students to data science concepts and guide them through hands-on projects. A teacher's guide is provided with solutions to the python challenges throughout the book.
- **Self-learners**: Individuals who are passionate about learning data analytics and machine learning outside of a formal classroom setting.

How to Use This Book

This book is structured to take students from foundational concepts in data analytics to hands-on projects in machine learning. Each chapter builds on the previous one, gradually increasing in complexity. Here's how you can get the most out of the material:

1. **Follow the Chapter Progression**: Start with the introductory chapters to understand the basics of data analytics, including data collection, cleaning, and visualization. From there, dive into machine learning concepts like supervised and unsupervised learning, neural networks, and model evaluation.

2. **Engage with Hands-on Activities**: Many chapters include practical exercises using real-world datasets from sources like Kaggle and Hugging Face. These exercises are designed to help you apply the concepts you've learned.

3. **Install the Right Tools**: To get the full experience, you'll need access to programming tools such as:

   - **Python**: The primary programming language used for data analytics in this book.
   - **Pandas and NumPy**: Libraries for data manipulation and analysis.
   - **Matplotlib and Seaborn**: Libraries for data visualization.
   - **Scikit-Learn**: A popular machine learning library.

Instructions for installing these tools and setting up a basic Python environment are included in **Appendix D**.

4. **Reflect on Real-World Applications**: Throughout the book, you'll find case studies that illustrate how data analytics and machine learning are used in real-world scenarios.

5. **Use the End-of-Chapter Questions**: Each chapter ends with review questions and practice problems to reinforce your understanding of the material.

Prerequisites

This book assumes no prior knowledge of data analytics or machine learning, making it accessible to beginners. However, the following skills or knowledge will help you succeed:

- **Basic Algebra**: Familiarity with concepts like variables, equations, and functions will help you understand the mathematical foundations of data analysis and machine learning algorithms.
- **Introductory Programming (Optional)**: While no advanced coding skills are required, having a basic understanding of programming concepts—like loops, variables, and functions— will be helpful when working through the coding exercises in Python.
- **Curiosity**: The most important prerequisite is a curiosity about data and how it shapes our world. A willingness to experiment and learn by doing will be your greatest asset.

Learning Objectives

This book is designed to help you achieve the following learning objectives:

- **Understand the Basics of Data Analytics**: By the end of the first few chapters, you'll be able to define data analytics, explain its importance, and describe how data is collected, cleaned, and analyzed.
- **Analyze Datasets with Python**: You'll learn to work with real-world datasets using Python, applying data manipulation techniques with libraries like Pandas and NumPy.
- **Visualize Data**: You'll explore how to create compelling data visualizations using tools like Matplotlib and Seaborn.
- **Apply Machine Learning Models**: As you progress through the book, you'll gain hands-on experience with machine learning algorithms, including regression, classification, and clustering techniques.
- **Evaluate and Improve Models**: You'll learn how to assess the performance of machine learning models and use techniques like cross-validation and hyperparameter tuning to improve them.
- **Work on Real-World Projects**: By the end of the book, you'll have completed several projects where you apply data analytics and machine learning to real-world problems using datasets from Kaggle and Hugging Face.

Chapter Features

Each chapter is designed to maximize your learning experience with the following features:

- **Key Terms**: Important vocabulary is highlighted throughout the chapters and defined in a glossary at the end of the book.
- **Case Studies**: Real-world case studies provide examples of how data analytics and machine learning are used in various industries, from healthcare to social media.

- **Practice Problems**: At the end of each chapter, you'll find review questions and practice problems that test your understanding of the material.
- **Hands-on Projects**: Every few chapters, you'll encounter larger projects that require you to apply what you've learned to real-world datasets. These projects are designed to simulate the kind of work data analysts and machine learning engineers do in the field.

Assessment and Review

To ensure that you're mastering the material, the book provides multiple ways to assess your understanding:

- **End-of-Chapter Questions**: After each chapter, you'll find questions that challenge you to reflect on the key concepts.
- **Hands-on Projects**: The projects are designed to be practical and comprehensive. They challenge you to apply multiple skills learned across chapters to solve real-world problems.
- **Solutions and Explanations**: For the review questions and practice problems, an answer key or solutions guide is available in a separate teacher's edition or online resource to help you check your work.

Additional Resources

To support your learning journey, here are some additional resources you may find useful:

- **Further Reading**: Throughout the book, you'll find recommendations for additional books, articles, and research papers that can deepen your understanding of data analytics and machine learning.
- **Online Courses**: Platforms like Coursera, edX, and Khan Academy offer free and paid courses on data analytics and machine learning. Consider enrolling in one of these courses for supplementary instruction.
- **Datasets**: In addition to the datasets used in this book, you can find a wealth of free datasets on platforms like Kaggle and Hugging Face, where you can practice your skills and experiment with different projects.

Appendix A: Glossary

A glossary at the end of the book provides definitions of key terms related to data analytics, machine learning, and statistics. It's a quick reference guide for students as they work through the book and encounter new terminology.

Appendix B: Python Kick Start

This appendix provides a concise introduction to Python, focusing on its application in data science and machine learning. It covers essential tools, libraries, and coding techniques, including data manipulation with pandas, numerical computations with NumPy, machine learning with scikit-learn, and neural network basics with

PyTorch. Designed for both beginners and experienced users, it offers a few practical examples to help readers quickly apply Python to machine learning tasks.

Appendix C: Data Sets and Source Code

An appendix at the end of the book provides a detailed list of key topics, terms, and datasets covered throughout the chapters, along with page references. This makes it easy for students to quickly find the information they need. It also provides information on how to obtain the source code and notebooks used in this book.

Appendix D: Installation and Requirements

This appendix provides step-by-step instructions for setting up the environment needed to run the notebooks included in the book. It outlines three methods for accessing and executing the notebooks—locally using Python and Anaconda, online through Google Colab, and on Kaggle's platform. The appendix also explains how to load datasets for exercises in each setup, ensuring readers can follow along with hands-on activities regardless of their preferred environment.

Chapter 1: What is Data Analytics?

Definition of Data Analytics

In today's world, data is everywhere. Whether it's the number of steps you take in a day, the purchases you make online, or the videos you watch on social media, data is being generated all the time. But data alone isn't very useful. It's what we do with data—how we analyze it—that makes it valuable. That's where *data analytics* comes in.

Data analytics is the process of examining raw data to uncover patterns, draw conclusions, and make informed decisions. Think of it as detective work, but instead of solving crimes, you're solving problems using numbers, facts, and trends hidden within data.

Data analytics helps us answer important questions, like:

- What products are customers most likely to buy?
- How can we improve a company's performance?
- What health trends are affecting different parts of the world?
- How do people interact with social media, and how can we improve their experience?

The goal of data analytics is to make sense of the data, extract useful information, and use that information to help solve real-world problems. Whether it's a company looking to improve sales or a scientist researching climate change, data analytics provides the tools to make smarter, evidence-based decisions.

Key Components of Data Analytics

Data analytics typically involves several steps:

1. **Collecting Data** – Gathering the raw information that you'll analyze. This could come from surveys, sensors, social media, or databases.
2. **Cleaning Data** – Data is rarely perfect, so it's important to remove or fix any errors. This might mean filling in missing data or correcting wrong information.
3. **Analyzing Data** – Once cleaned, the data is ready for analysis. Here, various techniques are used to find patterns, trends, and answers to questions.
4. **Interpreting Results** – After analysis, the results must be interpreted to understand what the data is telling us and to make decisions based on that insight.

In essence, data analytics transforms a large set of raw numbers into valuable insights that can be acted upon.

Types of Data Analytics

There are different types of data analytics, each serving a unique purpose:

- **Descriptive Analytics** answers the question, "What happened?" by summarizing past data. For example, a company might use descriptive analytics to look at last year's sales performance.

- **Diagnostic Analytics** goes one step further by asking, "Why did it happen?" This type of analysis helps identify causes of events or trends by digging deeper into the data.
- **Predictive Analytics** uses data from the past to predict future outcomes. For example, using data on customer purchases to predict what they'll buy next month.
- **Prescriptive Analytics** provides recommendations based on data, answering the question, "What should we do next?" This is where companies or organizations use data to make future decisions or solve a problem.

By understanding these core concepts of data analytics, you'll begin to see how powerful data can be, from improving a business to tackling global challenges. Data analytics opens the door to countless opportunities, and learning how to harness its power is key to making an impact in the world today.

Importance of Data in Today's World

In the 21st century, data is one of the most valuable resources in the world, often compared to oil for its transformative impact on society. Nearly every aspect of modern life is influenced by data, from the way we shop online to how we receive healthcare and even how governments make decisions. With the explosion of digital technologies, the amount of data generated every second is staggering—and it's growing every day.

Data Powers the Digital Economy

At the heart of today's digital economy is data. It drives the success of tech giants like Google, Amazon, and Facebook, and allows new industries like artificial intelligence, e-commerce, and financial technologies (FinTech) to thrive. Without data, these companies wouldn't be able to understand their customers, improve their products, or offer personalized experiences. For businesses large and small, data provides the insights they need to make better decisions, manage risks, and compete in an increasingly complex marketplace.

- **E-commerce** platforms like Amazon use data to recommend products tailored to your preferences, keeping you engaged and improving sales.
- **Streaming services** like Netflix rely on data to suggest movies and TV shows, based on your viewing habits and those of millions of other users.

The ability to harness and analyze vast amounts of data has given rise to new business models, where companies can innovate faster, solve problems more efficiently, and create personalized experiences for customers. This digital transformation would be impossible without data analytics.

Data Enables Scientific and Technological Breakthroughs

Beyond business, data is crucial in driving technological and scientific progress. In fields like healthcare, physics, and climate science, researchers are using data to make groundbreaking discoveries and solve some of the world's most pressing challenges.

- **Healthcare** is becoming more data-driven, with doctors using data from medical records, wearable devices, and genetic research to offer personalized treatments. This is transforming how we diagnose diseases and develop new medicines.
- **Climate scientists** rely on data to model future climate change scenarios, helping policymakers make decisions to protect the planet.

- **Technology companies** developing artificial intelligence (AI) and machine learning are feeding algorithms vast datasets to make systems smarter and more efficient. Self-driving cars, for example, use data from sensors to navigate roads safely.

These fields generate enormous amounts of data, and without the tools to analyze it, such advancements would be impossible. Data helps scientists understand complex systems and innovate solutions that impact millions of lives.

Data Shapes Our Daily Lives

You might not realize it, but data is deeply integrated into your everyday life. Every time you send a text, stream music, or order food online, data is being generated, collected, and analyzed. Even small things like using a fitness app to track your steps or receiving a weather forecast are powered by data.

- **Smart devices** in homes collect data on energy use, helping you save on electricity by suggesting more efficient habits.
- **GPS and navigation apps** use real-time data to give you the fastest route to your destination, taking into account traffic patterns and road conditions.

This constant flow of data has made life more convenient, but it also requires us to think about issues like privacy and data security. As consumers of technology, understanding how our data is used and protected is becoming increasingly important.

The Future is Data-Driven

The rise of data is not slowing down. In fact, the future will likely be even more data-driven than today. As more devices become connected to the internet (often referred to as the "Internet of Things" or IoT), the amount of data generated will multiply exponentially. Everything from your refrigerator to your car will collect and use data to make decisions.

- **Smart cities** will use data to optimize everything from traffic lights to waste management, improving the quality of life for citizens.
- **Education** will become more personalized, with data guiding how students learn and helping teachers tailor instruction to individual needs.
- **AI and automation** will continue to transform industries, making businesses more efficient and reducing the need for repetitive tasks.

Understanding data analytics is no longer a specialized skill—it's becoming a fundamental part of how the world operates. Whether you're looking at trends in the stock market or figuring out how to optimize your workout, data is the tool that unlocks new possibilities. By learning to analyze data, you gain the ability to understand the world more deeply and make more informed decisions about your future.

Applications of Data Analytics

Data analytics is used across a wide variety of industries and fields, helping organizations make more informed decisions, predict future trends, and optimize their operations. From business and healthcare to social media and sports, data analytics has transformed the way problems are solved and opportunities are identified. In this section, we'll explore some of the key applications of data analytics and how they're impacting different areas of our world.

1. Data Analytics in Business

One of the most widespread applications of data analytics is in business. Companies generate vast amounts of data from their daily operations, and analyzing this data allows them to improve efficiency, increase profits, and enhance customer experiences.

- **Customer Insights and Personalization**: Businesses use data analytics to better understand their customers' behaviors, preferences, and needs. By analyzing data from sales, website visits, and social media interactions, companies can predict what products a customer is likely to buy next and tailor marketing campaigns to individual users. For example, Amazon and Netflix use sophisticated algorithms to recommend products and shows based on customer data, boosting engagement and sales.
- **Optimizing Business Operations**: Data analytics is used to streamline business processes and improve decision-making. Companies can use data to optimize supply chain management, reduce waste, and improve inventory control. Retailers like Walmart use data analytics to forecast demand and ensure that the right products are available at the right time.
- **Financial Analysis and Risk Management**: In finance, data analytics is critical for assessing risk and identifying investment opportunities. Financial institutions use data analytics to detect fraudulent activities, analyze stock market trends, and assess credit risks. Hedge funds and trading firms rely heavily on data to make real-time investment decisions based on market data.

2. Data Analytics in Healthcare

The healthcare industry generates massive amounts of data from patient records, medical devices, and clinical research. Data analytics is playing an increasingly important role in improving patient care, reducing costs, and advancing medical research.

- **Predictive Medicine**: Using data analytics, healthcare providers can predict patient outcomes and tailor treatments accordingly. For instance, data from electronic health records (EHRs) can be analyzed to identify patients at risk of certain diseases, such as diabetes or heart disease, allowing doctors to intervene early and prevent serious health issues.
- **Medical Research and Drug Development**: Researchers use data analytics to accelerate drug discovery and medical research. By analyzing genetic data and clinical trial results, scientists can identify new treatments and develop personalized medicine tailored to individual patients' genetic profiles. During the COVID-19 pandemic, data analytics was essential in tracking the spread of the virus and developing vaccines.
- **Operational Efficiency**: Hospitals and clinics use data analytics to improve operational efficiency. For example, analyzing patient flow data can help healthcare facilities reduce wait times, allocate resources more effectively, and optimize scheduling for doctors and staff.

3. Data Analytics in Social Media

Social media platforms are vast ecosystems of data, and companies like Facebook, Twitter, and Instagram rely on data analytics to understand user behavior, deliver targeted ads, and manage content.

- **Content Personalization**: Social media platforms use data analytics to curate the content users see. Algorithms analyze what posts you interact with, who you follow, and what content you like or share to display content that is most likely to engage you. This keeps users on the platform longer and increases engagement.

- **Targeted Advertising**: Social media companies analyze user data to deliver personalized advertisements based on demographics, interests, and online behavior. This allows advertisers to target specific audiences more effectively, increasing the likelihood of conversions. For instance, if you frequently engage with posts about fitness, you're more likely to see ads for workout gear or health supplements.
- **Sentiment Analysis**: Companies and organizations use data analytics to perform sentiment analysis on social media, helping them understand public opinion on various topics. This is particularly useful for brands that want to monitor how their products or services are perceived by the public. During elections, politicians use social media data to gauge public sentiment on key issues and tailor their campaigns accordingly.

4. Data Analytics in Education

Data analytics is transforming the education sector by helping schools and universities improve student performance, enhance learning experiences, and streamline administrative processes.

- **Personalized Learning**: Schools use data analytics to track student progress and customize learning experiences. By analyzing data on student performance, teachers can identify areas where students are struggling and adapt their lessons to meet individual needs. This leads to more effective learning outcomes and ensures that no student is left behind.
- **Predicting Student Success**: Universities use data analytics to predict which students are at risk of dropping out or failing. By analyzing data on attendance, grades, and engagement, institutions can intervene early to provide support and improve student retention rates.

- **Administrative Decision-Making**: Data analytics helps educational institutions make better decisions regarding resource allocation, budgeting, and staffing. By analyzing enrollment trends and student feedback, schools can optimize class sizes, hire the right number of teachers, and allocate resources where they're most needed.

5. Data Analytics in Sports

In sports, data analytics is changing the way teams strategize, train, and compete. From player performance to game tactics, analytics has become a key factor in improving team outcomes and enhancing the fan experience.

- **Player Performance and Injury Prevention**: Teams use data analytics to monitor player performance and reduce the risk of injury. By analyzing data on player movements, heart rates, and fatigue levels, coaches can optimize training routines and rest periods to keep athletes in peak condition. In soccer, for instance, teams use wearable technology to track players' physical activity during games, providing insights on how to improve endurance and prevent injuries.
- **Game Strategy**: Data analytics is used to analyze opponents' tactics and develop game strategies. Coaches use data to identify strengths and weaknesses in opposing teams and make data-driven decisions during games. In basketball, the use of analytics to determine the most effective shots on the court has revolutionized the way the game is played, leading to the rise of three-point shooting.
- **Fan Engagement**: Sports teams also use data analytics to engage with fans and improve their experience. By analyzing ticket sales, social media interactions, and fan feedback, teams can create personalized promotions, improve stadium experiences, and grow their fan base.

6. Data Analytics in Government and Public Policy

Governments use data analytics to improve public services, make informed policy decisions, and address societal challenges such as crime, unemployment, and public health.

- **Policy Decision-Making**: Data analytics helps governments understand the impact of policies and make evidence-based decisions. For example, during the COVID-19 pandemic, governments used data analytics to track infection rates, predict hospital capacity needs, and determine when to implement lockdowns or vaccine rollouts.
- **Crime Prevention**: Law enforcement agencies use data analytics to predict and prevent crime. By analyzing crime patterns, police departments can allocate resources more effectively and develop strategies to reduce criminal activity. Predictive policing uses data to identify areas where crimes are more likely to occur, allowing law enforcement to be more proactive.
- **Urban Planning**: Data analytics plays a key role in urban planning and infrastructure development. Governments use data to analyze traffic patterns, public transportation usage, and population growth to plan new roads, schools, and public services. This helps cities grow sustainably and meet the needs of their citizens.

Conclusion

Data analytics has become a powerful tool that impacts nearly every sector of society. From helping businesses optimize operations to improving healthcare outcomes and even influencing public policy, data analytics is shaping the world we live in. Understanding how data analytics is applied across different industries provides a glimpse into its potential to drive innovation, solve problems, and create a better future.

Data-Driven Decision Making

In every aspect of life—whether it's in business, healthcare, education, or government—making the right decision is crucial. But how do people or organizations know they're making the right choice? Traditionally, decisions were made based on experience, intuition, or guesswork. However, in today's world, where data is abundant, decision-makers have a more reliable and powerful tool at their disposal: *data-driven decision making.*

Data-driven decision making (DDDM) refers to the practice of using data and analytics to guide decisions, rather than relying solely on intuition or anecdotal evidence. This approach involves collecting and analyzing relevant data to uncover trends, patterns, and insights that can help organizations or individuals make more informed and objective decisions.

Why Data-Driven Decision Making is Important

In an increasingly complex and fast-paced world, making decisions without solid evidence can lead to costly mistakes. Data-driven decision making offers several key advantages:

- **Accuracy and Objectivity**: Data provides a factual basis for decision-making. By relying on concrete numbers rather than gut feelings, decision-makers can avoid biases or assumptions that may lead to poor outcomes.

- **Efficiency**: Using data analytics allows organizations to process and analyze large amounts of information quickly. This leads to faster decision-making, especially in situations where time is of the essence.
- **Predictive Power**: One of the most powerful aspects of data-driven decision making is its ability to predict future outcomes. By analyzing past data, organizations can forecast trends, predict customer behavior, or anticipate potential risks. For example, businesses use predictive analytics to forecast demand for products or services, enabling them to prepare in advance and avoid shortages or overproduction.
- **Improved Accountability**: When decisions are backed by data, there's a clear record of why a particular choice was made. This transparency promotes accountability, as decisions can be traced back to the data that informed them.

How Data Drives Decisions in Different Sectors

1. **Business**:

   - Businesses use data to understand their customers, forecast market trends, and optimize operations. For instance, before launching a new product, companies analyze customer feedback, sales data, and market trends to determine whether there is enough demand. This helps them decide whether to invest in product development or adjust their strategy.
   - Data-driven marketing is another key area where businesses make decisions based on customer behavior, demographics, and purchasing patterns. For example, e-commerce companies use customer data to recommend products based on previous purchases, increasing the likelihood of sales.

2. **Healthcare**:

- In healthcare, data-driven decision making improves patient care by enabling doctors to make more informed diagnoses and treatment plans. For example, a hospital might analyze patient records to predict which individuals are at higher risk for certain conditions, such as diabetes or heart disease. This data allows doctors to recommend preventative measures, improving patient outcomes and reducing long-term healthcare costs.
- Data analytics can also assist in managing hospital resources, such as predicting patient admissions and optimizing staffing schedules to ensure adequate care during peak times.

3. **Education**:

- Schools and universities use data-driven decision making to improve student outcomes and optimize learning. By tracking student performance, attendance, and engagement, educators can identify students who may be at risk of falling behind. This data helps schools make decisions about resource allocation, such as providing additional tutoring or support for struggling students.
- On a larger scale, data analytics can inform decisions about curriculum design, helping educators understand which teaching methods are most effective.

4. **Government and Public Policy**:

- Governments use data to make informed decisions about public policy, resource allocation, and infrastructure planning. For example, during natural disasters, governments rely on data analytics to coordinate emergency responses, predict the impact of the event, and allocate resources effectively.

- In public health, data-driven decision making was critical during the COVID-19 pandemic, where governments analyzed infection rates, hospitalization data, and vaccination trends to make decisions about lockdowns, healthcare capacity, and vaccine distribution.

5. **Sports**:

- Sports teams rely heavily on data to make tactical decisions, improve player performance, and prevent injuries. Coaches and team managers analyze data from games, player statistics, and even real-time performance data during a match to make adjustments and optimize their strategies.
- Data also influences business decisions in sports, such as ticket pricing, merchandising, and fan engagement strategies.

Challenges in Data-Driven Decision Making

While data-driven decision making offers numerous advantages, it's not without its challenges. These include:

- **Data Quality**: For decisions to be accurate, the data must be reliable. Incomplete, outdated, or incorrect data can lead to flawed decisions. Ensuring data accuracy and consistency is a critical part of the decision-making process.
- **Data Overload**: With so much data available, it can be overwhelming to determine what's important and what's not. This challenge is often referred to as "data overload," and it requires effective data analysis tools and techniques to extract meaningful insights from large datasets.
- **Privacy Concerns**: With the increasing amount of data being collected, there are growing concerns about privacy and data

security. Organizations must ensure they handle sensitive data responsibly and comply with regulations, such as the General Data Protection Regulation (GDPR) or the California Consumer Privacy Act (CCPA).

Steps to Implementing Data-Driven Decision Making

To make data-driven decisions effectively, organizations need a clear process for collecting, analyzing, and acting on data. Here's a basic framework for implementing data-driven decision making:

1. **Identify the Problem or Question**: Start by defining the specific problem you need to solve or the question you need to answer. For example, a business might ask, "Why are sales declining in a particular region?"
2. **Collect Relevant Data**: Gather the data that will help answer your question. This could be sales records, customer feedback, social media trends, or operational data.
3. **Analyze the Data**: Use data analytics tools to identify patterns, trends, and insights. This step may involve data visualization, statistical analysis, or machine learning algorithms, depending on the complexity of the problem.
4. **Make the Decision**: Based on the insights gained from the data analysis, make your decision. For instance, if a company finds that customers in a certain region prefer a different product, they may decide to adjust their marketing strategy or product offerings.
5. **Evaluate the Outcome**: After implementing the decision, evaluate the results. Did the data-driven decision lead to the desired outcome? If not, you may need to revisit the data and adjust your approach.

Conclusion

Data-driven decision making is transforming how organizations operate, enabling them to make more informed, accurate, and efficient choices. By leveraging data, individuals and businesses can not only solve problems and predict future trends but also make decisions that are transparent, accountable, and backed by evidence. In a world where data is more accessible than ever, mastering data-driven decision making is essential for success.

Here are some reflective question to help you review what you learned in this chapter:

Reflective Questions

1. **What is Data Analytics?**

   - Define data analytics in your own words and explain why it is compared to detective work.

2. **Importance of Data Collection and Cleaning**

   - Why is the process of collecting and cleaning data critical before any actual data analysis takes place?

3. **Understanding Types of Analytics**

   - Describe each type of data analytics (Descriptive, Diagnostic, Predictive, and Prescriptive) and provide an example of a real-world application for each.

4. **Applications in Daily Life**

   - Can you think of an example from your daily life where data analytics could be used to improve an experience or service? Explain how data analytics could enhance this experience.

5. **Ethical Considerations**

   - What are some ethical considerations that should be taken into account when collecting and analyzing data?

6. **Future of Data Analytics**

   - Discuss how you think data analytics might change in the next ten years. What technologies could drive these changes?

7. **Role of Data in Decision Making**

   - Provide an example of how data-driven decision making can be more beneficial than intuition-based decision making in a business scenario.

8. **Impact of Poor Data Quality**

   - What are some potential consequences of making decisions based on poor quality data?

9. **Careers in Data Analytics**

   - Considering the skills discussed in this chapter, what type of careers do you think would be suitable for someone interested in data analytics?

10. **Personal Interaction with Data**

   - Reflect on how you interact with data in your everyday activities. How could an understanding of data analytics enhance your ability to make decisions or understand the world around you?

Chapter 2: Types of Data

Data is the foundation of data analytics and machine learning. But not all data is the same. Understanding the different types of data is crucial for analyzing it effectively and choosing the right tools for the job. In this chapter, we'll explore the two primary categories of data: **structured data** and **unstructured data**.

Structured Data

Structured data refers to data that is organized and easily searchable within a database or spreadsheet. This type of data is arranged in rows and columns, like tables, and follows a predefined format or schema. Structured data is often numerical or categorical and is commonly used in databases because of its ease of storage, retrieval, and analysis.

Examples of structured data include:

- **Spreadsheets** (e.g., Microsoft Excel or Google Sheets) where data is organized into rows and columns.
- **Relational Databases** (e.g., MySQL, Oracle) where data is stored in tables with relationships between different pieces of data.
- **Transactional Data** (e.g., sales records, inventory counts) which often include structured details like product names, prices, dates, and quantities.

Characteristics of Structured Data:

1. **Highly Organized**: Structured data follows a defined schema, making it easy to categorize and store in databases.
2. **Easily Searchable**: Because it is highly organized, searching through structured data is simple and can be done using database queries.
3. **Quantitative**: Structured data often contains numbers or categories that can be quantified, compared, and analyzed statistically.
4. **Fixed Fields**: Data is stored in fixed fields, with a specific meaning assigned to each field (e.g., "name," "date," "amount").

Examples of Structured Data in Practice:

- **Sales Data**: A company may use structured data to keep track of each transaction, including customer names, products purchased, and the total sales amount.
- **Healthcare Records**: Hospitals and clinics store structured data in the form of patient records, which may include a patient's age, gender, blood type, and medical history.
- **Weather Data**: Weather services store structured data on temperature, humidity, wind speed, and atmospheric pressure in a format that allows easy retrieval and analysis.

Benefits of Structured Data:

- **Efficient Storage**: Structured data can be stored efficiently in relational databases, where relationships between data points can be easily managed.
- **Quick to Analyze**: Data analysts and machine learning models can quickly analyze structured data to generate insights because the data follows a consistent format.

- **Strong Data Integrity**: Because structured data must adhere to a schema, it's easier to maintain the integrity of the data and ensure accuracy.

Limitations of Structured Data:

- **Limited Flexibility**: Structured data must fit within a predefined schema, which can limit its ability to capture complex information.
- **Inability to Handle Unstructured Information**: Structured data is not ideal for representing more nuanced or qualitative information, such as text, images, or videos.

Unstructured Data

While structured data is neat and organized, **unstructured data** is the opposite. Unstructured data refers to data that does not have a predefined structure or schema, and therefore, cannot be easily organized into rows and columns. This type of data is typically qualitative in nature, such as images, videos, emails, or social media posts.

Unstructured data is more difficult to analyze using traditional tools, but it holds vast amounts of valuable information that can be unlocked using advanced analytics, machine learning, or natural language processing (NLP).

Examples of unstructured data include:

- **Text Documents** (e.g., Word documents, PDF files)
- **Social Media Posts** (e.g., tweets, Facebook status updates)

- **Multimedia Files** (e.g., images, audio recordings, video files)
- **Emails** (e.g., email conversations that contain text, attachments, and metadata)

Characteristics of Unstructured Data:

1. **Lack of Format**: Unstructured data does not follow a fixed format, making it more challenging to store and analyze compared to structured data.
2. **Qualitative**: Unstructured data is often qualitative, such as text or media files, rather than numerical or categorical.
3. **Large Volume**: Unstructured data is typically produced in large quantities, especially with the growth of social media, video streaming, and digital communication.
4. **Requires Advanced Tools**: Analyzing unstructured data often requires specialized tools and techniques, such as machine learning, to extract useful insights.

Examples of Unstructured Data in Practice:

- **Social Media Analytics**: Businesses often analyze unstructured data from social media to gain insights into customer sentiment. For example, Twitter posts or Facebook comments can be analyzed using sentiment analysis tools to determine how customers feel about a product.
- **Customer Support Emails**: Customer service teams may process unstructured data from emails to track common complaints, identify keywords, and address customer issues more effectively.
- **Video and Image Processing**: Companies like YouTube and Google process vast amounts of unstructured data in the form of videos and images. Machine learning algorithms can be used to classify images or recommend videos based on user preferences.

Benefits of Unstructured Data:

- **Rich Information**: Unstructured data can contain a wealth of valuable information that is not captured by traditional structured datasets. This includes opinions, ideas, and other qualitative insights.
- **More Flexibility**: Since unstructured data isn't limited to a schema, it can capture a wide variety of information, including emotional tone, visual content, and more.
- **Scalability**: With the right tools, organizations can process massive amounts of unstructured data, opening up new opportunities for insights.

Limitations of Unstructured Data:

- **Difficult to Store**: Unstructured data cannot be stored in traditional relational databases. Instead, it requires alternative storage solutions like NoSQL databases.
- **Challenging to Analyze**: Extracting meaningful insights from unstructured data often requires sophisticated machine learning models and natural language processing techniques.
- **Time-Consuming**: Processing and analyzing unstructured data can be more time-consuming and computationally intensive than structured data due to its complexity.

Comparison of Structured and Unstructured Data

Feature	Structured Data	Unstructured Data
Organization	Follows a fixed schema; stored in rows and columns	No predefined structure or organization
Data Type	Mostly numerical or categorical (e.g., dates, product names)	Text, images, videos, emails, social media posts
Storage	Stored in relational databases (e.g., SQL)	Stored in NoSQL databases or data lakes
Ease of Analysis	Easier to analyze with traditional tools like Excel or SQL	Requires advanced tools such as machine learning and NLP
Volume	Generally smaller in size	Often much larger due to multimedia and social media data
Example	Sales data, patient records, financial transactions	Social media posts, video files, customer support emails

Conclusion

Understanding the differences between structured and unstructured data is fundamental to working with data analytics and machine learning. Structured data is highly organized and easily processed with traditional analytical tools, but it is limited in its ability to capture complex or qualitative information. Unstructured data, while more difficult to manage and analyze, holds the potential for richer, deeper insights when processed using advanced techniques.

Now that we've explored the distinction between structured and unstructured data, it's important to understand another key classification of data: **quantitative** vs. **qualitative** data. These two types of data differ in how they represent information and the kind of insights they can offer. Whether the data is structured or unstructured, it can fall into one of these two categories, each of which plays a crucial role in data analytics.

Quantitative vs Qualitative Data

When working with data, it's essential to understand how the information is expressed. This brings us to two major types of data: **quantitative data** and **qualitative data**. These types are often used hand-in-hand to create a full picture of whatever is being studied, but they serve different purposes and are used for different kinds of analysis.

Quantitative Data

Quantitative data refers to any information that can be measured and expressed in numerical terms. This type of data is often used to quantify variables, meaning it provides concrete, measurable values that can be analyzed using mathematical and statistical methods. Quantitative data is essential for performing calculations, creating graphs, and building predictive models in machine learning.

Characteristics of Quantitative Data:

1. **Numerical**: Quantitative data is always expressed as numbers (e.g., quantities, percentages, rates).
2. **Objective**: It provides objective measurements, meaning the data is not influenced by personal feelings or interpretations.
3. **Discrete or Continuous**: Quantitative data can be discrete (finite, countable values) or continuous (infinite, measurable values that can take any value within a range).

   - **Discrete Data**: This type of quantitative data represents countable items (e.g., the number of students in a class, the number of cars sold).
   - **Continuous Data**: Continuous data represents measurements on a continuous scale (e.g., height, weight, time, temperature).

Examples of Quantitative Data:

- **Sales Figures**: The total revenue generated by a company in a specific quarter (e.g., $50,000 in Q3).
- **Test Scores**: A student's score on an exam (e.g., 85 out of 100).
- **Temperature Data**: The daily high temperatures recorded over a week (e.g., 72°F, 75°F, 80°F).
- **Weight of Objects**: The weight of products in a manufacturing facility (e.g., 12.5 kg, 18.3 kg).

Benefits of Quantitative Data:

- **Precision**: Quantitative data provides precise, measurable values that can be used for detailed statistical analysis.

- **Comparability**: Since it is numerical, quantitative data can be compared across different datasets or over time.
- **Predictive Power**: In machine learning, quantitative data is often used to build predictive models, as numbers can be processed and analyzed by algorithms effectively.

Limitations of Quantitative Data:

- **Limited Context**: While quantitative data provides precise measurements, it may not offer much context or explanation behind those numbers. For example, sales figures alone cannot explain why sales increased or decreased without further investigation.
- **Not Suitable for Complex Concepts**: Certain concepts, such as emotions, opinions, or customer satisfaction, are difficult to capture using purely quantitative measures.

Qualitative Data

Qualitative data refers to non-numerical information that describes qualities, characteristics, or attributes of a subject. Unlike quantitative data, which focuses on "how much" or "how many," qualitative data answers questions like "what," "how," and "why." It is often subjective and more descriptive, providing insights that are harder to quantify but essential for understanding deeper meanings and context.

Characteristics of Qualitative Data:

1. **Descriptive**: Qualitative data is expressed in words, symbols, or categories, rather than numbers.

2. **Subjective**: It often reflects opinions, perceptions, or experiences, making it more subjective and open to interpretation.
3. **Categorical**: Qualitative data is typically divided into categories or classifications that describe different attributes (e.g., colors, types, or labels).

Examples of Qualitative Data:

- **Customer Reviews**: Written feedback from customers describing their experiences with a product (e.g., "The product is durable but a bit too expensive").
- **Interview Transcripts**: Transcriptions of interviews where participants describe their thoughts or experiences (e.g., "I enjoy using the app because it's user-friendly").
- **Social Media Comments**: Posts and comments on platforms like Twitter or Instagram (e.g., "This restaurant has amazing service!").
- **Survey Responses**: Open-ended responses to survey questions (e.g., "I would recommend this service because it saved me time").

Benefits of Qualitative Data:

- **Rich Context**: Qualitative data provides deeper insights into the reasoning, motivations, and experiences behind certain behaviors or events. For example, a customer review can explain why a person is satisfied or dissatisfied with a product.
- **Flexibility**: It captures complex phenomena that cannot be easily reduced to numbers, such as emotions, opinions, or social dynamics.
- **Broad Insights**: Qualitative data is useful when exploring new topics or generating ideas, as it allows for open-ended exploration.

Limitations of Qualitative Data:

- **Difficult to Analyze**: Since qualitative data is non-numerical, it can be challenging to analyze systematically. It often requires more manual interpretation or advanced techniques such as natural language processing (NLP).
- **Subjectivity**: Because it is often based on opinions or perceptions, qualitative data can be influenced by personal biases and may not always be reliable for objective analysis.
- **Hard to Generalize**: Unlike quantitative data, which can be easily generalized across larger populations, qualitative data may be specific to individual cases or small samples, making it harder to draw broad conclusions.

Comparison of Quantitative and Qualitative Data

While both quantitative and qualitative data provide valuable insights, they serve different purposes. Here's a quick comparison to highlight their key differences:

Feature	Quantitative Data	Qualitative Data
Nature	Numerical	Descriptive
Example	Sales figures, test scores, temperatures	Customer reviews, interview responses, social media posts
Analysis Method	Statistical and mathematical analysis	Thematic analysis, content analysis

Feature	Quantitative Data	Qualitative Data
Data Format	Numbers, percentages, rates	Words, images, symbols
Precision	High precision, exact values	Provides broad, contextual insights
Use Case	Best for measuring quantities or trends	Best for exploring experiences, opinions, and perceptions

Blending Quantitative and Qualitative Data

In many cases, both quantitative and qualitative data are used together to provide a fuller understanding of the problem at hand. For example:

- **In Business:** A company may analyze **quantitative data** like customer purchase frequency, but also gather **qualitative data** through customer reviews or surveys to understand *why* certain products perform better than others.
- **In Healthcare:** Doctors use **quantitative data** such as blood pressure readings or lab results, but they also rely on **qualitative data** such as patient feedback or medical history to create personalized treatment plans.
- **In Education:** Schools may use **quantitative data** such as test scores to assess student performance, but **qualitative data** from teacher observations or student feedback provides a more comprehensive view of a student's progress.

Conclusion

Both quantitative and qualitative data are crucial to data analytics, each providing unique insights into a problem. Quantitative data offers measurable and precise information, ideal for statistical analysis and prediction. Qualitative data, on the other hand, offers depth and context, giving meaning to the numbers. By understanding the differences and applications of both types of data, you can approach data analysis more holistically, gaining a more complete understanding of the subjects you study.

In the next section, we'll dive deeper into how data is collected, processed, and prepared for analysis, and explore real-world datasets that combine both quantitative and qualitative elements.

Now that we've explored the differences between quantitative and qualitative data, the next crucial step in the data analytics process is understanding how this data is gathered. The way data is collected plays a major role in determining its quality and relevance to the problem you're trying to solve. In this section, we will look at some common **data collection methods** used in both structured and unstructured data scenarios, as well as best practices to ensure the data is reliable and meaningful.

Data Collection Methods

Collecting data is one of the foundational steps in any data analytics process. The methods used to collect data will vary depending on whether you are working with quantitative or qualitative data, as well as whether the data is structured or unstructured. Good data collection ensures that you gather accurate, complete, and relevant information that will later be used for analysis.

Let's explore the most common data collection methods, how they work, and when they are used.

1. Surveys and Questionnaires

Surveys and questionnaires are widely used to collect both quantitative and qualitative data. They involve asking a set of predefined questions to a group of respondents to gather information on a particular topic.

- **Quantitative Data**: When designed with closed-ended questions (e.g., multiple-choice, yes/no, or rating scales), surveys can gather structured, quantitative data. This type of data is easy to quantify and analyze statistically. For example, a customer satisfaction survey asking, "On a scale of 1-10, how satisfied are you with our service?" will yield numerical data.
- **Qualitative Data**: Surveys can also be designed to collect qualitative data through open-ended questions (e.g., "What do you like most about our product?"). These responses are often longer, written text that provides deeper insights into customer opinions or experiences.

Benefits:

- **Scalable**: Surveys can reach a large audience quickly, especially through online platforms.
- **Versatile**: They can collect both quantitative and qualitative data depending on the question format.

- **Response Bias**: The quality of data can be influenced by how questions are phrased or the respondent's willingness to answer honestly.
- **Limited Depth**: While closed-ended questions provide structured data, they may not capture the full range of insights compared to more open-ended, qualitative methods.

2. Observations

Observation involves collecting data by watching and recording behaviors, actions, or events in their natural settings. It's especially useful for collecting **qualitative data** and can be structured or unstructured depending on the level of control the observer has over the environment.

- **Structured Observation**: In this method, the observer follows a predetermined plan, watching for specific behaviors or events. The data collected is often structured and quantitative. For example, observing how many customers enter a store within an hour and counting specific behaviors like purchases or inquiries.
- **Unstructured Observation**: This method involves a more flexible, qualitative approach, where the observer notes behaviors, interactions, or reactions without a strict checklist. For example, an anthropologist studying social interactions within a community.

Benefits:

- **Natural Behavior**: Observations can provide insights into real-world behavior without interference from external questions or prompts.

- **Contextual Data**: Observations often capture context and details that surveys or interviews may miss, especially in qualitative research.

Challenges:

- **Time-Consuming**: Observational studies can take a long time and often require the observer to be physically present.
- **Observer Bias**: The observer's own perceptions or biases may affect how behaviors are recorded or interpreted.

3. Interviews

Interviews are a direct data collection method where one person (the interviewer) asks another person (the interviewee) a series of questions. Interviews can be structured (with a strict set of questions) or unstructured (more open and exploratory). They are typically used to collect **qualitative data**, providing deep insights into thoughts, opinions, and experiences.

- **Structured Interviews**: These follow a set list of questions, often producing data that can be more easily compared across different respondents.
- **Unstructured Interviews**: These are more flexible and open-ended, allowing the interviewee to explore different topics and provide richer qualitative data. For example, asking broad questions like "Can you describe your experience using our product?" without predetermined answer options.

Benefits:

- **Rich, In-Depth Data**: Interviews allow for a deeper exploration of complex issues, emotions, and motivations.

- **Flexibility**: Unstructured interviews can adapt to the flow of conversation, leading to unexpected insights.

Challenges:

- **Resource-Intensive**: Interviews can be time-consuming and expensive, especially if conducted in person.
- **Subjective**: The quality and depth of data collected depend heavily on the skills of the interviewer and the honesty of the interviewee.

4. Transactional Data Collection

Transactional data refers to data generated by systems or devices during regular operations, such as sales transactions, customer purchases, or website visits. This data is typically **structured** and **quantitative**, making it ideal for analysis.

- **Examples**: Point-of-sale systems, e-commerce platforms, and online banking generate transactional data, such as the total amount spent, time of purchase, items bought, and payment method used.

Benefits:

- **Automated and Accurate**: Transactional data is automatically recorded by systems, minimizing the risk of human error and providing high-quality data.
- **Real-Time**: Many systems capture data in real time, allowing for up-to-the-minute analysis.

Challenges:

- **Limited Context**: While transactional data is rich in numbers, it lacks the qualitative context to explain *why* certain transactions happen.
- **Privacy Concerns**: The collection and use of personal transaction data must comply with data privacy regulations, such as GDPR or CCPA.

5. Social Media and Web Scraping

Social media platforms and websites generate vast amounts of **unstructured data** in the form of posts, comments, images, and user interactions. Web scraping is the process of automatically collecting data from websites, which can then be analyzed for insights.

- **Social Media**: Platforms like Twitter, Facebook, and Instagram provide an abundance of unstructured qualitative data, such as user opinions, reviews, or posts that reflect public sentiment on various topics.
- **Web Scraping**: This technique is used to extract data from websites, which can range from structured (e.g., product listings, pricing information) to unstructured (e.g., blog posts, forum discussions).

Benefits:

- **Massive Data Volume**: Social media and web scraping provide access to large, often real-time data sets that reflect a wide range of opinions and trends.
- **Cost-Effective**: Web scraping can collect large volumes of data with relatively low cost, compared to manual data collection methods.

Challenges:

- **Unstructured and Noisy**: Social media data is often unstructured and can contain irrelevant or noisy information that needs to be filtered out.
- **Legal and Ethical Concerns**: Web scraping may violate website terms of service, and social media data collection must comply with privacy policies and ethical guidelines.

6. Sensors and IoT Devices

With the rise of the Internet of Things (IoT), devices and sensors are increasingly used to collect **quantitative** data from the physical world. These devices monitor and record data on temperature, humidity, motion, or even human vitals, providing precise, real-time data streams.

- **Examples**: Smart home devices, environmental sensors, and health monitoring wearables like Fitbits or Apple Watches generate vast amounts of structured quantitative data.

Benefits:

- **Real-Time Monitoring**: Sensors provide continuous, real-time data collection, enabling immediate analysis and action.
- **Automation**: Data is collected automatically, minimizing human intervention and error.

Challenges:

- **Data Overload**: The sheer volume of data generated by IoT devices can be overwhelming and require advanced data storage and processing techniques.

- **Data Integration**: Combining data from different sensors and devices can be complex, especially when the data formats are not standardized.

7. Experimental Data Collection

In experimental data collection, researchers design controlled experiments to collect **quantitative** data. These experiments manipulate one or more variables to observe their effect on a specific outcome, often using randomization to ensure fairness.

- **Example**: A scientist may design an experiment to test the effect of a new drug on patients by dividing them into two groups—one receiving the drug and the other receiving a placebo—and measuring outcomes such as blood pressure.

Benefits:

- **Controlled Variables**: Experimental data provides precise measurements and allows for the testing of cause-and-effect relationships.

Challenges:

- **Artificial Environment**: Experiments may not reflect real-world conditions, leading to data that is not entirely generalizable.

Conclusion

The method you choose to collect data will depend on the type of data you need—whether it's quantitative, qualitative, structured, or unstructured. Each method has its advantages and challenges, and selecting the right one ensures that the data is reliable, relevant, and actionable for your analysis.

Now that we've covered various methods of data collection, the next step is understanding how datasets are structured and used in data analytics and machine learning. Whether data is collected through surveys, sensors, or social media, it must be organized into datasets that can be analyzed effectively. In this section, we will introduce the concept of datasets, how they are used in analytics, and explore popular platforms like Kaggle and Hugging Face, which provide access to a vast array of datasets for learning and experimentation.

Basic Introduction to Datasets

A **dataset** is a collection of related data organized for analysis, typically structured in rows and columns like a table. Each row represents an individual **instance** (a specific data point), and each column represents a **feature** (a variable or attribute of the data). Datasets are essential in both data analytics and machine learning because they serve as the foundation for developing insights, training models, and making predictions.

The format and structure of a dataset can vary significantly depending on the type of data (structured, unstructured, quantitative, qualitative) and the specific problem it is designed to solve. In some cases, datasets consist of numerical values like sales figures or measurements, while in others, they might include text, images, or audio files. Regardless of the data type, working with well-organized and clean datasets is critical for successful analysis.

Popular Platforms for Datasets: Kaggle and Hugging Face

There are many places to find datasets for practicing data analysis and machine learning. Two of the most popular platforms are **Kaggle** and **Hugging Face**. These platforms offer access to vast arrays of datasets, ranging from structured tables to unstructured text, images, and audio files, making them invaluable resources for both beginners and advanced learners.

1. Kaggle

Kaggle is a widely used platform in the data science community. It not only provides datasets but also serves as a hub for competitions, learning resources, and collaboration with other data professionals.

Key Features of Kaggle:

- **Wide Variety of Datasets**: Kaggle hosts datasets across many fields, such as business, healthcare, finance, and sports. Famous examples include the **Titanic Dataset** for classification and the **House Prices Dataset** for regression.
- **Competitions**: Kaggle is known for its data science competitions, where users solve real-world problems using provided datasets and compete for prizes.
- **Community Notebooks**: Kaggle users often share their analysis and machine learning projects through public notebooks, which can serve as helpful learning tools.

- **Built-in Tools**: Kaggle offers a free cloud-based environment where you can code in Python, perform data analysis, and build machine learning models directly on the platform.

Example Datasets on Kaggle:

- **Titanic Dataset**: This dataset includes demographic and travel information about the passengers aboard the Titanic. It is commonly used to teach beginners how to predict survival rates using classification models.
- **House Prices Dataset**: This dataset contains various features of homes (e.g., size, number of rooms) and their sale prices, making it ideal for practicing regression techniques.

2. Hugging Face

Hugging Face is a platform best known for its work in **natural language processing (NLP)**, though it also offers datasets for other machine learning tasks such as image and audio processing. Hugging Face provides access to numerous datasets, many of which are specifically designed for tasks that involve unstructured data like text or speech.

Key Features of Hugging Face:

- **Specialization in NLP**: Hugging Face is the go-to platform for NLP tasks, offering datasets and pre-trained models for language translation, text summarization, sentiment analysis, and more.
- **Dataset Hub**: Hugging Face hosts a variety of datasets, from text-heavy data for NLP to image and audio datasets for more advanced machine learning tasks.
- **Pre-trained Models**: Hugging Face is also well known for its library of pre-trained machine learning models, particularly

in the **Transformers** library, which can be fine-tuned for different tasks.

Example Datasets on Hugging Face:

- **IMDb Movie Reviews**: This dataset contains movie reviews labeled as positive or negative, making it perfect for sentiment analysis.
- **Common Voice**: An open-source dataset from Mozilla containing voice recordings, often used for speech recognition tasks.

Best Practices for Working with Datasets

Regardless of whether you're working with structured or unstructured data, or using datasets from Kaggle or Hugging Face, there are several best practices to follow to ensure that you maximize the value of your analysis:

1. **Understand the Dataset**: Before diving into analysis, take time to explore the dataset's structure, including its features and target variable. Understand what each column represents and identify any potential issues like missing values.
2. **Data Cleaning**: Data cleaning is a crucial step to ensure the accuracy of your results. This involves handling missing data, removing duplicates, and fixing any errors.
3. **Exploratory Data Analysis (EDA)**: Use EDA techniques such as histograms, scatter plots, and summary statistics to understand patterns and relationships in the data.
4. **Respect Privacy and Ethics**: Be mindful of data privacy and ethical concerns, especially when working with personal or sensitive data. Always comply with regulations such as GDPR and CCPA.

5. **Start Small**: When working with large datasets, it's helpful to begin with a smaller sample to get a feel for the data before performing full-scale analysis or training a machine learning model.

Conclusion

Datasets are the core components of data analytics and machine learning. Platforms like **Kaggle** and **Hugging Face** provide easy access to diverse datasets, allowing learners and professionals alike to practice and refine their skills. Whether you're working with structured datasets for predictive modeling or unstructured text for natural language processing, understanding and properly managing your datasets is key to successful analysis.

Now that we've introduced the concept of datasets and explored popular platforms like Kaggle, it's time to put this knowledge into practice with a real-world dataset. One of the most famous datasets for learning data analytics and machine learning is the **Titanic dataset** on Kaggle. This dataset provides a great opportunity for beginners to practice key skills like data cleaning, exploration, and building predictive models.

In this case study, we'll dive into the **Titanic dataset**, guide you through understanding the data, and get hands-on with analyzing it. By the end of this case study, you'll have a strong foundation in handling real-world datasets, and you'll start building your own machine learning models.

Case Study: The Titanic Dataset

The **Titanic dataset** is one of the most iconic datasets in data science. It contains information about the passengers on the RMS Titanic, which tragically sank in 1912 after hitting an iceberg. The dataset includes various features like age, gender, ticket class, and whether or not the passenger survived. The goal of this case study is to explore the dataset, clean it up, and use it to predict which passengers survived the disaster.

Step 1: Accessing the Titanic Dataset

The first step is to access the dataset. Kaggle provides a simple interface for downloading and exploring datasets. Here's how you can access the Titanic dataset:

1. **Visit Kaggle**: Go to the Kaggle website (www.kaggle.com) and search for the "Titanic" dataset or follow this direct link: Titanic - Machine Learning from Disaster[1].
2. **Download the Dataset**: On the dataset page, download the **train.csv** and **test.csv** files. These are the two main files you'll work with.
3. **Upload to Your Environment**: If you're not using a local machine and instead using Kaggle's notebook environment, upload these files so they're ready for analysis.

Step 2: Understanding the Titanic Dataset

Before analyzing the dataset, let's take a closer look at the features (columns) it contains. Understanding the structure of the data is essential for effective analysis:

- **PassengerId**: A unique identifier for each passenger.

---

[1]https://www.kaggle.com/c/titanic

- **Survived**: This is the target variable (0 = No, 1 = Yes) indicating whether the passenger survived or not.
- **Pclass**: The passenger's class (1st, 2nd, 3rd), a proxy for socioeconomic status.
- **Name**: The name of the passenger.
- **Sex**: The gender of the passenger.
- **Age**: The age of the passenger.
- **SibSp**: The number of siblings or spouses aboard the Titanic.
- **Parch**: The number of parents or children aboard the Titanic.
- **Ticket**: The passenger's ticket number.
- **Fare**: The fare paid by the passenger.
- **Cabin**: The cabin number (if available).
- **Embarked**: The port where the passenger boarded the Titanic (C = Cherbourg; Q = Queenstown; S = Southampton).

Step 3: Exploring the Titanic Dataset (EDA)

Exploratory Data Analysis (EDA) is a crucial step in understanding your dataset. In this step, you'll use Python and libraries like Pandas, Matplotlib, and Seaborn to uncover trends and relationships in the data.

Let's begin by loading the dataset and exploring its structure:

```
import pandas as pd

# Load the Titanic dataset
train_data = pd.read_csv('train.csv')

# Display the first few rows of the dataset
train_data.head()
```

Output

	PassengerId	Survived	Pclass	Name	Sex	Age	SibSp	Parch	Ticket	Fare	Cabin	Embarked
0	1	0	3	Braund, Mr. Owen Harris	male	22.0	1	0	A/5 21171	7.2500	NaN	S
1	2	1	1	Cumings, Mrs. John Bradley (Florence Briggs Th...	female	38.0	1	0	PC 17599	71.2833	C85	C
2	3	1	3	Heikkinen, Miss. Laina	female	26.0	0	0	STON/O2. 3101282	7.9250	NaN	S
3	4	1	1	Futrelle, Mrs. Jacques Heath (Lily May Peel)	female	35.0	1	0	113803	53.1000	C123	S
4	5	0	3	Allen, Mr. William Henry	male	35.0	0	0	373450	8.0500	NaN	S

Figure 1. The First 5 Rows of Titanic Data

This will give you an overview of the dataset and help you understand what kind of data you're working with.

Basic Dataset Insights:

- **Check for missing data**: Some features, like "Age" and
  "Cabin," might have missing values. We need to identify these
  and decide how to handle them.

```
1   # Check for missing values
2   print(train_data.isnull().sum())
```

Output PassengerId 0 Survived 0 Pclass 0 Name 0 Sex 0 Age 177
SibSp 0 Parch 0 Ticket 0 Fare 0 Cabin 687 Embarked 2

Observations Looking at the output, it seems that Age, Cabin, and
Embarked all have missing data. Later in this chapter we'll learn
how to deal with these gaps.

- **Statistical Summary**: Let's generate a summary of the
  dataset to understand the distribution of numerical features
  like age, fare, and survival rate.

```
1   # Summary statistics
2   train_data.describe()
```

	PassengerId	Survived	Pclass	Age	SibSp	Parch	Fare
count	891.000000	891.000000	891.000000	714.000000	891.000000	891.000000	891.000000
mean	446.000000	0.383838	2.308642	29.699118	0.523008	0.381594	32.204208
std	257.353842	0.486592	0.836071	14.526497	1.102743	0.806057	49.693429
min	1.000000	0.000000	1.000000	0.420000	0.000000	0.000000	0.000000
25%	223.500000	0.000000	2.000000	20.125000	0.000000	0.000000	7.910400
50%	446.000000	0.000000	3.000000	28.000000	0.000000	0.000000	14.454200
75%	668.500000	1.000000	3.000000	38.000000	1.000000	0.000000	31.000000
max	891.000000	1.000000	3.000000	80.000000	8.000000	6.000000	512.329200

Figure 2. Statistical Summary of the Titanic Data

Here are a couple of interesting observations from these statistics:

Survival Rate: The mean of the "Survived" column is approximately 0.38, indicating that only about 38% of passengers survived. This low survival rate could spark curiosity about which factors (e.g., age, class) might have influenced a passenger's chances.

Fare Range: The "Fare" column shows a huge range, with prices going from 0 to over 500. This might indicate differences in class and wealth among passengers, possibly hinting at how social class could have impacted survival rates.

- **Distribution of Survivors**: One of the first questions you'll want to explore is how many passengers survived versus how many didn't. Let's visualize this distribution:

```
1    import seaborn as sns
2    import matplotlib.pyplot as plt
3
4    # Plot the distribution of survivors
5    sns.countplot(x='Survived', data=train_data)
6    plt.title('Survival Distribution')
7    plt.show()
```

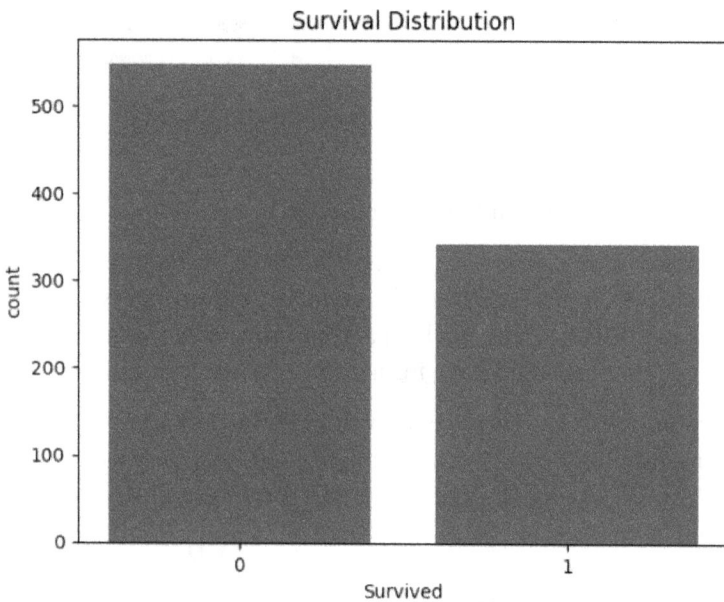

Figure 3. Survivors vs. Non-Survivors

Exploratory Questions:

- How does **gender** affect survival?
- Does the **passenger class (Pclass)** correlate with survival?
- What is the impact of **age** on survival?

Step 4: Data Cleaning and Preprocessing

In the Titanic dataset, we will likely encounter missing or inconsistent data that needs to be addressed before we can perform any meaningful analysis or build a machine learning model.

Handling Missing Values:

- **Age**: The "Age" column has missing values. One common approach is to fill in missing ages with the median or mean age of the passengers.

```
# Fill missing values in 'Age' with the median age
train_data['Age'].fillna(train_data['Age']
                  .median(), inplace=True)
```

- **Embarked**: The "Embarked" column also has missing values. Since there are only two missing entries, we can fill them with the most frequent port of embarkation.

```
# Fill missing values in 'Embarked' with the most
# frequent value
train_data['Embarked'].fillna
    (train_data['Embarked'].mode()[0],
    inplace=True)
```

- **Cabin**: The "Cabin" feature has many missing values, so we might choose to drop it for simplicity, especially for a beginner case study.

```
1   # Drop the 'Cabin' column
2   train_data.drop(columns=['Cabin'], inplace=True)
```

Step 5: Feature Engineering

Feature engineering involves creating new features or modifying existing ones to improve the performance of a machine learning model. For example:

- **Family Size**: Instead of dealing with "SibSp" and "Parch" separately, you can create a new feature "FamilySize" by adding the two together.

```
1   # Create a new feature 'FamilySize'
2   train_data['FamilySize'] = train_data['SibSp'] +
3       train_data['Parch']
```

- **Gender Conversion**: The "Sex" column needs to be converted into numerical format for the model. Let's convert "male" to 0 and "female" to 1.

```
1   # Convert 'Sex' to numerical format
2   train_data['Sex'] = train_data['Sex'].map
3       ({'male': 0, 'female': 1})
```

Step 6: Building a Simple Predictive Model

Now that the data has been cleaned and processed, you can start building a simple predictive model using a machine learning algorithm such as Logistic Regression. Logistic regression is a statistical method used for binary classification that predicts the probability of an event's occurrence by fitting data to a logistic curve. It is particularly useful in scenarios where you need to classify outcomes into two distinct categories, such as predicting whether an email is spam or not spam, or in this case, predicting whether someone survived the Titanic or not.

```python
from sklearn.model_selection import
    train_test_split
from sklearn.linear_model import
    LogisticRegression
from sklearn.metrics import accuracy_score

# Define features and target variable
features = ['Pclass', 'Sex', 'Age', 'Fare',
    'FamilySize']
X = train_data[features]
y = train_data['Survived']

# Split the data into training and validation sets
X_train, X_val, y_train, y_val = train_test_split
    (X, y, test_size=0.2, random_state=42)

# Create and train the model
model = LogisticRegression()
model.fit(X_train, y_train)

# Make predictions and evaluate accuracy
y_pred = model.predict(X_val)
accuracy = accuracy_score(y_val, y_pred)
print(f'Accuracy of the model: {accuracy *
    100:.2f}%')
```

Output Accuracy of the model: 80.45%

Step 7: Evaluating and Improving the Model

After building your first model, you can evaluate its performance and experiment with different algorithms, such as Decision Trees or Random Forests, to improve accuracy. You can also try **hyperparameter tuning** or use **cross-validation** for more robust performance.

Here is a list of algorithms used for predicting classification:

Here's a table of machine learning algorithms similar to Logistic Regression, often used for classifying data like that from the Titanic dataset. Each entry includes a description, as well as pros and cons:

Algorithm	Description	Pros	Cons
Logistic Regression	A statistical model that estimates the probability of a binary outcome.	Simple, efficient, interpretable.	Assumes linear relationships, sensitive to outliers.
Decision Trees	Models that predict the value of a target variable by learning decision rules from features.	Easy to understand and interpret, non-linear.	Can easily overfit, sensitive to data variance.

Algorithm	Description	Pros	Cons
Random Forest	An ensemble of Decision Trees, typically trained via the bagging method.	Robust, handles overfitting better than Decision Trees.	Can be complex, slow to predict.
Support Vector Machines (SVM)	Separates data with a hyper-plane, maximiz-ing the margin between classes.	Effective in high-dimensional spaces.	Requires scaling, not suitable for large datasets.
K-Nearest Neighbors (KNN)	Classifies new cases based on a similarity measure (e.g., distance functions).	No training involved, simple.	Computationally expensive, sensitive to imbalanced data.
Naive Bayes	Applies Bayes' theorem with the assumption of independence between features.	Fast, works well with high dimensions, robust to irrelevant features.	Assumes feature independence, poor estimates of probabilities.

Algorithm	Description	Pros	Cons
Gradient Boosting	Boosts weak learners (typically decision trees) sequentially, correcting errors.	Often provides predictive accuracy that cannot be beat.	Prone to overfitting, requires careful tuning.

Each of these algorithms can be implemented in Python, using the `scikit-learn` library. When choosing an algorithm, consider the nature of your data and the specific requirements of your problem, such as the need for model interpretability or prediction speed.

Conclusion

This case study introduced you to the basics of working with real-world datasets by using the Titanic dataset. You learned how to load and explore the dataset, clean and preprocess the data, and build a simple predictive model. While this is just the beginning, this experience provides a strong foundation for working with more complex datasets and machine learning techniques in the future.

Now that you've completed the Titanic case study, you've seen firsthand how real-world datasets often come with challenges such as missing values, inconsistent formats, or irrelevant features. To build effective models and draw meaningful insights, it's essential to clean and preprocess the data before diving into analysis or machine learning.

In the next chapter, we'll dive deeper into data cleaning and preprocessing, starting with one of the most common issues: handling missing values. By mastering these techniques, you'll be able to prepare any dataset for accurate analysis, ensuring your results are both reliable and actionable.

Here are 10 questions that reflect the content of the chapter along with a few simple Python challenges to help you apply what you've learned:

Reflective Questions

1. **Define Structured and Unstructured Data:**

   - What is structured data and what makes it different from unstructured data?

2. **Examples Identification:**

   - Give three examples of structured data and three examples of unstructured data from your daily life.

3. **Characteristics Comparison:**

   - List at least two characteristics each of structured and unstructured data.

4. **Importance of Data Types:**

   - Why is it important to distinguish between structured and unstructured data when performing data analytics?

5. **Benefits and Limitations:**

   - Discuss one benefit and one limitation of using structured data.

6. **Unstructured Data Challenges:**

- What are some of the challenges associated with analyzing unstructured data, and how can they be overcome?

7. **Real-world Applications:**

   - Describe a real-world application where unstructured data might provide more insights than structured data.

8. **Tools for Data Analysis:**

   - What types of tools or technologies would you use to analyze a large dataset of unstructured social media posts?

9. **Impact on Decision Making:**

   - How can the choice between using structured or unstructured data impact business decision-making?

10. **Future of Data Analytics:**

    - Predict how the role of unstructured data in data analytics might evolve over the next decade.

Python Challenges

1. **Loading Data:**

   - Write a Python script to load the titanic train.csv file into a pandas DataFrame and display the first five rows. Assume the file is named 'data.csv'.

2. **Data Inspection:**

- Use Python to check for missing values in the DataFrame loaded above.

3. **Basic Data Analysis:**

   - Write a Python function that calculates the mean of a column in a DataFrame. Use this function to find the mean of a numerical column in 'train.csv'.

4. **Data Visualization:**

   - Create a histogram to visualize the distribution of a numerical column in 'train.csv' using matplotlib.

5. **Handling Missing Data:**

   - Write a Python script to fill missing values in a column of your DataFrame with the column's median value.

Chapter 3: Data Cleaning and Preprocessing

Handling Missing Values

One of the most common challenges in working with real-world datasets is dealing with **missing values**. Missing data can arise from various sources, such as human error, incomplete data entry, system failures, or simply because certain information was unavailable. Regardless of the cause, missing values can negatively affect your analysis and the performance of machine learning models, so handling them effectively is essential.

In this chapter, we'll explore different strategies for handling missing values and provide practical examples using Python.

1. Identifying Missing Values

The first step in handling missing data is **identifying** where the missing values are located in your dataset. In Python, you can use the Pandas library to detect missing values quickly.

Let's take a simple example using the **Titanic dataset**, which contains missing values in features like **Age** and **Cabin**.

```
1   import pandas as pd
2
3   # Load the Titanic dataset
4   train_data = pd.read_csv('train.csv')
5
6   # Display the first few rows of the dataset
7   print(train_data.head())
8
9   # Check for missing values
10  print(train_data.isnull().sum())
```

The .isnull().sum() function returns the number of missing values in each column. For example, the output might show something like this:

```
1   PassengerId        0
2   Survived           0
3   Pclass             0
4   Name               0
5   Sex                0
6   Age              177
7   SibSp              0
8   Parch              0
9   Ticket             0
10  Fare               0
11  Cabin            687
12  Embarked           2
13  dtype: int64
```

From the output, we can see that:

- **Age** has 177 missing values.
- **Cabin** has 687 missing values.
- **Embarked** has 2 missing values.

2. Assessing the Impact of Missing Values

Before deciding how to handle missing values, it's important to assess how much missing data is present and whether the missing values are random or exhibit some pattern. Missing values can be classified into three categories:

1. **Missing Completely at Random (MCAR)**: The missingness is independent of any variable in the dataset (e.g., random technical errors).
2. **Missing at Random (MAR)**: The missingness is related to some observed data (e.g., younger passengers are more likely to have missing values for "Cabin").
3. **Missing Not at Random (MNAR)**: The missingness is related to the value itself (e.g., wealthier passengers who paid higher fares might be less likely to have missing cabin information due to privacy).

To assess the percentage of missing values in each feature, you can use the following code:

```
# Calculate the percentage of missing values in
# each column
missing_percentage = train_data.isnull().mean() *
    100
print(missing_percentage)
```

3. Common Strategies for Handling Missing Values

Once you've identified the missing data and assessed its extent, you can choose an appropriate method for handling it. Below are some common techniques, with examples of how to implement them in Python.

3.1 Removing Missing Values

One straightforward approach is to **remove rows or columns** that contain missing values. This is a viable option when the number of missing values is small, and removing them won't significantly affect the analysis.

- **Dropping Rows**: If a few rows have missing values, you can remove them using the dropna() function.

```
1   # Drop rows with missing values
2   train_data_cleaned = train_data.dropna()
3   print(train_data_cleaned.isnull().sum())
```

- **Dropping Columns**: If a column has a large number of missing values, like the **Cabin** feature, and it's not essential to your analysis, you can drop the entire column.

```
1   # Drop the 'Cabin' column due to excessive
2   # missing values
3   train_data = train_data.drop(columns=['Cabin'])
```

When to use this method:

- If only a small percentage of the data is missing (e.g., less than 5-10% of the dataset).
- If the column with missing data is irrelevant to the analysis or the target variable.

3.2 Imputing Missing Values

Instead of removing data, you can **impute** or replace missing values with an estimated value. Common imputation methods include using the **mean**, **median**, or **mode** of the existing data. This is particularly useful for numerical features like **Age** in the Titanic dataset.

- **Mean Imputation**: Replace missing values with the mean of the column.

```
# Fill missing 'Age' values with the mean age
train_data['Age'].fillna(train_data['Age'].mean()
    , inplace=True)
```

- **Median Imputation**: Replace missing values with the median of the column, which is useful when the data is skewed.

```
# Fill missing 'Age' values with the median age
train_data['Age'].fillna(train_data['Age']
                .median(), inplace=True)
```

- **Mode Imputation**: For categorical data, you can replace missing values with the most frequent value (mode). For example, the **Embarked** column can be filled this way:

```
1  # Fill missing 'Embarked' values with the most
2  # common port of embarkation
3  train_data['Embarked'].fillna
4      (train_data['Embarked'].mode()[0],
5      inplace=True)
```

When to use this method:

- When missing values make up a moderate proportion of the data.
- For numerical data, median imputation is better than mean when there are outliers.
- For categorical data, mode imputation is commonly used.

3.3 Advanced Imputation Techniques

For more sophisticated datasets, you might want to use **advanced imputation techniques** like **interpolation** or **model-based imputation**.

- **Interpolation:** This technique uses patterns in the data to estimate missing values. It's particularly useful when the data is time-series or has a logical progression (e.g., missing temperatures over time).

```
1  # Interpolating missing values in a time series
2  # or continuous data
3  train_data['Age'] = train_data['Age']
4                      .interpolate()
```

- **Model-Based Imputation:** You can train a machine learning model to predict the missing values based on other features. For example, you can use a regression model to predict missing ages based on passenger class, fare, and gender.

Note: We'll first need to create Sex into a number regression can work with:

```
# Convert 'Sex' column to numeric
df['Sex'] = df['Sex'].map({'male': 0, 'female':
    1})
```

Now let's predict the missing Age values and add them back into our dataset

```
from sklearn.linear_model import LinearRegression

# Separate the data into rows with and without
# missing 'Age'
train_data_with_age =
    train_data[train_data['Age'].notnull()]
train_data_without_age =
    train_data[train_data['Age'].isnull()]

# Select relevant features for prediction
features = ['Pclass', 'Sex', 'Fare', 'SibSp',
    'Parch']

# Create a simple regression model to predict
# 'Age'
X = train_data_with_age[features]
y = train_data_with_age['Age']

model = LinearRegression()
model.fit(X, y)

# Predict missing 'Age' values
train_data_without_age['Age'] = model.predict
    (train_data_without_age[features])

```

```
26   # Combine the datasets
27   train_data = pd.concat([train_data_with_age,
28       train_data_without_age])
```

When to use this method:

- When you want to capture more complex relationships in the data.
- When missing data is not random and can be predicted based on other available features.

4. Evaluating the Impact of Missing Value Handling

After handling missing values, it's important to evaluate whether the choices you made have affected the quality of the dataset. You can do this by:

- **Re-checking for missing values**: Ensure that no missing values remain, unless they are intentional or unavoidable.

```
1   print(train_data.isnull().sum())
```

- **Analyzing changes in distribution**: For numerical data, check how imputation affected the distribution of the feature. For example, if you imputed ages using the median, visualize the new distribution:

```
1    import seaborn as sns
2    import matplotlib.pyplot as plt
3
4    # Visualize the distribution of 'Age' before and
5    # after imputation
6    sns.histplot(train_data['Age'], kde=True)
7    plt.title('Age Distribution After Imputation')
8    plt.show()
```

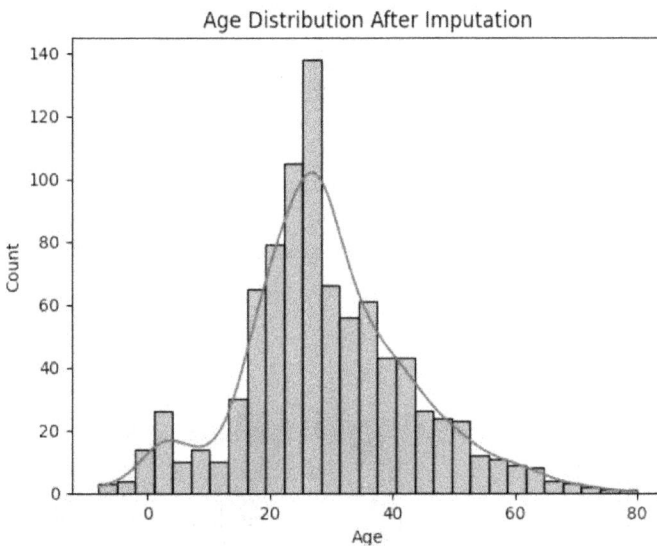

Figure 4. Age distribution after Imputation

- **Testing Model Performance**: If you're building a machine learning model, compare how different missing value handling techniques (e.g., dropping vs. imputing) impact model accuracy.

5. Conclusion

Handling missing values is an essential step in preparing a dataset for analysis. Whether you choose to drop rows, impute missing values, or use more advanced techniques like model-based imputation, the key is to ensure that your data is clean and reliable before proceeding to the next stages of your analysis or machine learning workflow.

Now that you've learned how to handle missing values, another common issue that can affect the quality of your dataset is **duplicate records**. Duplicates can distort analysis and mislead models, making your results inaccurate. Identifying and removing duplicate entries is an important part of the data cleaning process, ensuring that your dataset reflects the true distribution of information.

In this section, we'll explore how to detect and remove duplicates, and provide practical examples to help you clean your dataset efficiently.

Removing Duplicates

Duplicate records occur when the same data point or entry appears multiple times in your dataset. This can happen for various reasons, such as data entry errors, system glitches, or redundant data collection. While duplicates may seem harmless, they can lead to overestimating certain trends or relationships in your analysis. In machine learning, duplicates can bias model training, resulting in overfitting or skewed predictions.

1. Identifying Duplicates

Before removing duplicates, you first need to identify where duplicates exist in your dataset. Pandas provides simple functions for checking whether your data contains any duplicate rows.

Let's use the **Titanic dataset** again for this example:

```
1  import pandas as pd
2
3  # Load the Titanic dataset
4  train_data = pd.read_csv('train.csv')
5
6  # Check for duplicate rows
7  duplicates = train_data.duplicated()
8  print(f"Number of duplicate rows:
9         {duplicates.sum()}")
```

The duplicated() function returns a boolean Series where True indicates a duplicate row. Summing this Series gives the total number of duplicates in the dataset.

2. Removing Duplicate Rows

Once you've identified duplicates, you can remove them from your dataset using the drop_duplicates() function. By default, this function removes only rows that are **completely identical** in all columns.

```
1  # Remove duplicate rows
2  train_data_cleaned = train_data.drop_duplicates()
3
4  # Check if duplicates are removed
5  print(f"Number of duplicate rows after removal:
6      {train_data_cleaned.duplicated().sum()}")
```

Example Output:

```
1   Number of duplicate rows: 2
2   Number of duplicate rows after removal: 0
```

In this example, any rows that were exact duplicates have been removed. The dataset is now free of duplicate entries, ensuring that each row represents a unique data point.

3. Removing Duplicates Based on Specific Columns

Sometimes, you may want to remove duplicates based on certain columns rather than the entire dataset. For example, in the Titanic dataset, you might only care about whether a passenger is duplicated based on their **Name** or **Ticket Number**, even if other features like **Cabin** or **Fare** differ slightly.

You can specify columns when removing duplicates:

```
1   # Remove duplicates based on the 'Name' and
2   # 'Ticket' columns
3   train_data_cleaned = train_data.drop_duplicates
4       (subset=['Name', 'Ticket'])
5
6   # Check for duplicates based on specific columns
7   print(f"Number of duplicates after removing based
8       on 'Name' and 'Ticket':
9       {train_data_cleaned.duplicated(subset=['Name',
10          'Ticket']).sum()}")
```

This method ensures that only rows with identical values in the specified columns (in this case, Name and Ticket) are considered duplicates and removed.

4. Keeping the First or Last Occurrence of Duplicates

By default, drop_duplicates() keeps the **first occurrence** of a duplicate and removes the rest. However, in some cases, you might want to keep the **last occurrence** of the duplicate instead. You can control this behavior using the keep parameter.

- **Keeping the first occurrence** (default behavior):

```
1   # Keep the first occurrence of each duplicate
2   train_data_cleaned = train_data.drop_duplicates
3      (keep='first')
```

- **Keeping the last occurrence**:

```
1   # Keep the last occurrence of each duplicate
2   train_data_cleaned = train_data.drop_duplicates
3      (keep='last')
```

5. Removing Duplicate Columns

Sometimes, datasets contain duplicate **columns** as well. For instance, you may have multiple columns with the same data due to a merging error or redundant data entry.

To identify and remove duplicate columns, you can compare the values in each column:

```
1   # Transpose the dataset and drop duplicate columns
2   train_data_cleaned = train_data.T.drop_duplicates
3      ().T
```

This method transposes the dataset (so rows become columns), removes any duplicate columns, and then transposes it back.

6. Best Practices for Removing Duplicates

- **Understand the Cause**: Before removing duplicates, it's important to understand why they exist. In some cases, duplicates may not need to be removed if they represent legitimate data (e.g., two passengers with the same name but different tickets).
- **Check for Partial Duplicates**: In some cases, rows may have duplicate values in critical columns, but differences in other columns. Always review the impact of removing duplicates on other aspects of the data.
- **Back Up Your Data**: When removing duplicates, always work on a copy of the dataset to avoid losing important data unintentionally.

7. Evaluating the Impact of Removing Duplicates

Once you've removed duplicates, evaluate the dataset to ensure its integrity. Here are a few checks you can perform:

- **Check for unexpected drops** in the size of the dataset.
- **Visualize key metrics** before and after removing duplicates to ensure that no important trends or patterns were lost.

For example, you might check how removing duplicates impacts the distribution of survival rates in the Titanic dataset:

```
1   import seaborn as sns
2   import matplotlib.pyplot as plt
3
4   # Compare the survival rate before and after
5   # removing duplicates
6   sns.countplot(x='Survived',
7       data=train_data_cleaned)
8   plt.title('Survival Distribution After Removing \
9       Duplicates')
10  plt.show()
```

Conclusion

Removing duplicates is a simple yet crucial step in the data cleaning process. Duplicates can distort analysis and lead to misleading results in machine learning models, so identifying and handling them properly is essential. Whether you're removing exact duplicates or filtering based on specific columns, the techniques outlined in this section will help you ensure that your dataset is clean and ready for accurate analysis.

Now that we've covered how to clean your dataset by handling missing values and removing duplicates, the next step in the data preprocessing pipeline is **data transformation**. One of the most important aspects of data transformation is ensuring that your numerical features are on the same scale, so that they are properly understood by machine learning algorithms.

In this section, we'll explore two critical techniques for transforming numerical data: **Normalization** and **Standardization**. Both of these techniques help improve the performance and accuracy of machine learning models by ensuring that different features are comparable and contribute equally to the model.

Normalization and Standardization

When working with datasets, especially those with numerical features like age, salary, or temperature, you'll often encounter variables that have very different ranges. For instance, in the Titanic dataset, the **Fare** ranges from 0 to 500, while **Age** ranges from 0 to around 80. If left untransformed, these differences can negatively impact certain machine learning algorithms, which assume that all features contribute equally to the model. This is where **Normalization** and **Standardization** come into play.

1. Normalization

Normalization (also known as **min-max scaling**) transforms your data so that it falls within a specific range, typically between 0 and 1. This technique adjusts the scale of the features while preserving their relative relationships. It is particularly useful when you have a dataset where the values of features vary widely but you want to ensure that each feature has equal importance during analysis.

Formula for Normalization:

$$X_{\text{normalized}} = \frac{X - X_{\text{min}}}{X_{\text{max}} - X_{\text{min}}}$$ Where:

- X is the original value.
- X_{min} is the minimum value of the feature.
- X_{max} is the maximum value of the feature.

When to Use Normalization:

Normalization is generally recommended when:

- The algorithm you're using doesn't assume a normal distribution of data (e.g., **K-Nearest Neighbors, Neural Networks**).
- You need the data to fall within a specific range, particularly for gradient-based optimization methods (such as Neural Networks).

Example: Applying Normalization in Python

Let's apply normalization to the **Fare** and **Age** columns of the Titanic dataset using **scikit-learn**'s `MinMaxScaler`.

```
1   from sklearn.preprocessing import MinMaxScaler
2
3   # Select the features to normalize
4   features_to_normalize = ['Fare', 'Age']
5
6   # Initialize the MinMaxScaler
7   scaler = MinMaxScaler()
8
9   # Apply normalization
10  train_data[features_to_normalize] =
11      scaler.fit_transform
12      (train_data[features_to_normalize])
13
14  # Check the result
15  print(train_data[['Fare', 'Age']].head())
```

Output | | Fare | Age | |—|————-|————-| | 0 | 0.014151 | 0.271174 | | 1 | 0.139136 | 0.472229 | | 2 | 0.015469 | 0.321438 | | 3 | 0.103644 | 0.434531 | | 4 | 0.015713 | 0.434531 |

After applying normalization, the values in the **Fare** and **Age** columns will be scaled to fall between 0 and 1.

2. Standardization

Standardization (also called **z-score normalization**) transforms data so that it has a mean of 0 and a standard deviation of 1. This method is particularly useful when the data has different scales but follows a **normal distribution** (bell curve). Standardization ensures that each feature contributes equally to the model, regardless of the original range of values.

Formula for Standardization:

$$X_{\text{standardized}} = \frac{X - \mu}{\sigma}$$ Where:

- (X) is the original value.
- (μ) is the mean of the feature.
- (σ) is the standard deviation of the feature.

When to Use Standardization:

Standardization is commonly used when:

- The algorithm assumes that the data is normally distributed (e.g., **Logistic Regression, Support Vector Machines, Linear Regression**).
- The model is sensitive to the magnitude of features (such as **Principal Component Analysis (PCA)** or **K-Means Clustering**).

Example: Applying Standardization in Python

Let's apply standardization to the **Fare** and **Age** columns using **scikit-learn**'s `StandardScaler`.

```
1    from sklearn.preprocessing import StandardScaler
2
3    # Select the Features to Standardize
4    features_to_standardize = ['Fare', 'Age']
5
6    # Initialize the StandardScaler
7    scaler = StandardScaler()
8
9    # Apply standardization
10   train_data[features_to_standardize] =
11       scaler.fit_transform
12       (train_data[features_to_standardize])
13
14   # Check the result
15   print(train_data[['Fare', 'Age']].head())
```

Output | | Fare | Age | |—|————-|————-| | 0 | -0.502445 | -0.565736 | | 1 | 0.786845 | 0.663861 | | 2 | -0.488854 | -0.258337 | | 3 | 0.420730 | 0.433312 | | 4 | -0.486337 | 0.433312 |

After standardization, the values in the **Fare** and **Age** columns will be centered around 0, with most values falling between -1 and 1.

3. Comparison of Normalization and Standardization

Aspect	Normalization (Min-Max Scaling)	Standardization (Z-Score)
Output Range	Typically between 0 and 1	Mean of 0 and standard deviation of 1
Use Cases	Algorithms like KNN, Neural Networks	Algorithms like SVM, Logistic Regression, PCA

Aspect	Normalization (Min-Max Scaling)	Standardization (Z-Score)
Effect on Data	Preserves relationships, scales data proportionally	Centers data around 0, standardizes variance
Assumption of Distribution	No assumption about the distribution of the data	Assumes data is approximately normally distributed

4. When to Use Each Method

- **Use Normalization** when you know that the algorithm you're using benefits from having all input features in the same range, particularly when features vary significantly in scale. This is common in deep learning and K-Nearest Neighbors (KNN).
- **Use Standardization** when you know the data follows a normal distribution and when your algorithm assumes or benefits from data with a mean of 0 and a standard deviation of 1. This is essential in algorithms like Support Vector Machines (SVM) or Linear Regression.

5. Evaluating the Impact of Normalization and Standardization

After normalizing or standardizing your data, it's important to evaluate how the transformation affects your dataset. You can visualize the distribution of features before and after scaling to see the changes.

For example, you can plot the distribution of the **Fare** and **Age** features before and after normalization or standardization:

```python
import seaborn as sns
import matplotlib.pyplot as plt

# Plot the distribution of 'Fare' and 'Age' after
# transformation
fig, ax = plt.subplots(1, 2, figsize=(12, 6))

sns.histplot(train_data['Fare'], kde=True,
    ax=ax[0])
ax[0].set_title('Fare Distribution After
    Transformation')

sns.histplot(train_data['Age'], kde=True,
    ax=ax[1])
ax[1].set_title('Age Distribution After
    Transformation')

plt.show()
```

Output

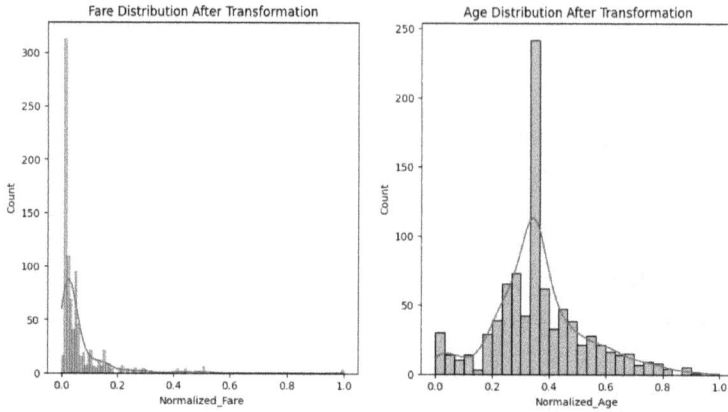

Figure 5. Feature Distribution after Normalization

Conclusion

Both **Normalization** and **Standardization** are essential techniques in data preprocessing that help ensure your numerical features are scaled appropriately for machine learning models. Depending on the type of algorithm and the nature of your data, one method may be more suitable than the other. By applying these techniques, you can improve the performance and accuracy of your models and ensure that all features contribute equally to the final outcome.

In the next section, we'll discuss **encoding categorical variables**, which is another crucial step in transforming your dataset to make it machine-learning ready.

Now that we've covered the essential techniques of normalization and standardization, the next step in mastering data cleaning and preprocessing is understanding the tools and libraries that make these tasks easier. Two of the most important Python libraries for working with data are **Pandas** and **NumPy**. These libraries provide efficient ways to manipulate, clean, and transform datasets, making them indispensable for anyone working in data analytics or machine learning.

In this chapter, we'll introduce you to **Pandas** and **NumPy**, focusing on how these libraries can be used for cleaning and preprocessing data, ensuring your datasets are ready for analysis and modeling.

Introduction to Pandas and NumPy

When it comes to working with datasets, Python's **Pandas** and **NumPy** libraries offer powerful tools to perform a wide variety of data manipulation and preprocessing tasks. Whether you need to clean up messy data, handle missing values, or prepare your dataset for machine learning, these libraries are the backbone of any data science workflow.

1. Overview of Pandas

Pandas is a high-level data manipulation library that provides data structures and functions for efficiently manipulating datasets. The two primary data structures in Pandas are **DataFrames** (for handling tabular data) and **Series** (for handling one-dimensional data). Pandas excels at data wrangling, such as cleaning data, transforming it into different formats, and preparing it for analysis.

Key Features of Pandas:

- **DataFrames**: Pandas' primary structure for handling datasets. A DataFrame is similar to a table or spreadsheet with rows and columns.
- **Handling Missing Data**: Pandas provides functions for identifying, filling, or removing missing values.
- **Data Aggregation and Grouping**: You can group data by categories and aggregate statistics such as sums, means, and counts.
- **Merging and Joining**: Combine multiple datasets using merge, join, or concatenate functions.
- **Filtering and Sorting**: Pandas makes it easy to filter rows based on conditions and sort data.

Let's explore some core Pandas functionalities, focusing on how they can be used for cleaning and preprocessing data.

Example 1: Loading and Inspecting Data

```
import pandas as pd

# Load a dataset into a DataFrame
df = pd.read_csv('train.csv')

# Display the first few rows of the DataFrame
print(df.head())

# Display basic information about the dataset
print(df.info())
```

This will load the dataset into a Pandas DataFrame, allowing you to inspect its structure, columns, and data types.

Example 2: Handling Missing Data with Pandas

As already discusssed, Pandas provides simple functions to handle missing data, such as filling or dropping missing values.

```
1   # Check for missing values
2   print(df.isnull().sum())
3
4   # Fill missing 'Age' values with the median
5   df['Age'].fillna(df['Age'].median(), inplace=True)
6
7   # Drop rows where 'Embarked' is missing
8   df.dropna(subset=['Embarked'], inplace=True)
9
10  # Verify missing values are handled
11  print(df.isnull().sum())
```

With these functions, you can efficiently clean up your dataset, ensuring that it's free of missing values before proceeding with further analysis or model building.

Example 3: Filtering and Sorting Data

Pandas also allows you to filter and sort your dataset based on specific conditions.

```
1   # Filter passengers who survived
2   survived_passengers = df[df['Survived'] == 1]
3
4   # Sort passengers by 'Fare' in descending order
5   sorted_by_fare = df.sort_values(by='Fare',
6       ascending=False)
7   sorted_by_fare.head()
```

Output

	PassengerId	Survived	Pclass	Name	Sex	Age	SibSp	Parch	Ticket	Fare	Cabin	Embarked
258	259	1	1	Ward, Miss. Anna	female	35.0	0	0	PC 17755	512.3292	NaN	C
737	738	1	1	Lesurer, Mr. Gustave J	male	35.0	0	0	PC 17755	512.3292	B101	C
679	680	1	1	Cardeza, Mr. Thomas Drake Martinez	male	36.0	0	1	PC 17755	512.3292	B51 B53 B55	C
88	89	1	1	Fortune, Miss. Mabel Helen	female	23.0	3	2	19950	263.0000	C23 C25 C27	S
27	28	0	1	Fortune, Mr. Charles Alexander	male	19.0	3	2	19950	263.0000	C23 C25 C27	S

Figure 6. Fare in Descending order of survivors

These basic operations are useful for inspecting and manipulating data during the cleaning process, allowing you to gain insights before moving to more complex transformations.

---

2. Overview of NumPy

NumPy is a fundamental library for performing numerical operations in Python. It provides support for large multi-dimensional arrays and matrices, along with a variety of mathematical functions to operate on these arrays efficiently. While Pandas is designed for higher-level data manipulation, NumPy is the go-to library for performing mathematical and statistical operations, especially when working with numerical data.

Key Features of NumPy:

- **Efficient Array Manipulation**: NumPy's **ndarray** object allows for fast and efficient manipulation of large datasets.
- **Element-wise Operations**: You can apply mathematical operations across entire arrays.
- **Linear Algebra and Statistical Functions**: NumPy includes a variety of functions for linear algebra, statistical analysis, and random number generation.

While Pandas is often used for handling tabular data, NumPy is used for more granular numerical operations.

Example 1: Creating and Manipulating NumPy Arrays

NumPy arrays can be created from lists or directly using NumPy functions.

```
1   import numpy as np
2
3   # Create a NumPy array from a list
4   array = np.array([1, 2, 3, 4, 5])
5
6   # Perform element-wise operations
7   array_squared = array ** 2
8   print(f'array squared {array_squared}')
9
10  # Create a 2D array (matrix)
11  matrix = np.array([[1, 2], [3, 4], [5, 6]])
12
13  # Access elements
14  print(matrix[0, 1])   # Access element at row 0,
15      column 1
```

Output array squared [1 4 9 16 25] 2

Arrays in NumPy allow you to efficiently perform mathematical operations on large datasets, making it a perfect tool for working with numerical data.

Example 2: Using NumPy for Standardization and Normalization

As we just discovered, NumPy is often used in conjunction with Pandas to perform normalization or standardization across a dataset. For example, you can use NumPy's mathematical functions to apply transformations to numerical features.

- **Standardization** with NumPy:

```
1   # Standardize a feature (Age)
2   age_mean = np.mean(df['Age'])
3   age_std = np.std(df['Age'])
4
5   df['Age_standardized'] =
6       (df['Age'] - age_mean)/age_std
```

Output

	PassengerId	Survived	Pclass	Name	Sex	Age	SibSp	Parch	Ticket	Fare	Cabin	Embarked	Age_standardized
0	1	0	3	Braund, Mr. Owen Harris	male	22.0	1	0	A/5 21171	7.2500	NaN	S	-0.530377
1	2	1	1	Cumings, Mrs. John Bradley (Florence Briggs Th.	female	38.0	1	0	PC 17599	71.2833	C85	C	0.571831
2	3	1	3	Heikkinen, Miss. Laina	female	26.0	0	0	STON/O2 3101282	7.9250	NaN	S	-0.254825
3	4	1	1	Futrelle, Mrs. Jacques Heath (Lily May Peel)	female	35.0	1	0	113803	53.1000	C123	S	0.365167
4	5	0	3	Allen, Mr. William Henry	male	35.0	0	0	373450	8.0500	NaN	S	0.366167

Figure 7. Output from Standardizing Age

- **Normalization** with NumPy:

```
1   # Normalize a feature (Fare)
2   fare_min = np.min(df['Fare'])
3   fare_max = np.max(df['Fare'])
4
5   df['Fare_normalized'] =
6       (df['Fare'] - fare_min)/(fare_max - fare_min)
```

Output

	PassengerId	Survived	Pclass	Name	Sex	Age	SibSp	Parch	Ticket	Fare	Cabin	Embarked	Age_standardized	Fare_normalized
0	1	0	3	Braund, Mr. Owen Harris	male	22.0	1	0	A/5 21171	7.2500	NaN	S	-0.530377	0.014151
1	2	1	1	Cumings, Mrs. John Bradley (Florence Briggs Th...	female	38.0	1	0	PC 17599	71.2833	C85	C	0.571831	0.139136
2	3	1	3	Heikkinen, Miss. Laina	female	26.0	0	0	STON/O2 3101282	7.9250	NaN	S	-0.254825	0.015469
3	4	1	1	Futrelle, Mrs. Jacques Heath (Lily May Peel)	female	35.0	1	0	113803	53.1000	C123	S	0.366167	0.103644
4	5	0	3	Allen, Mr. William Henry	male	35.0	0	0	373450	8.0500	NaN	S	0.366167	0.015713

Figure 8. Output after Normalizing Fare

Here, NumPy provides the mathematical backbone for these transformations, while Pandas helps in managing the dataset structure.

---

3. Integrating Pandas and NumPy for Data Preprocessing

When cleaning and preprocessing datasets, Pandas and NumPy often work together to handle the entire pipeline from loading data to transforming it for machine learning models. NumPy's array-based computation is highly efficient for performing mathematical transformations, while Pandas provides a high-level structure for managing the dataset.

Example: Complete Data Preprocessing Workflow with Pandas and NumPy

Let's walk through a full basic data preprocessing workflow using both Pandas and NumPy. This will involve handling missing values, scaling numerical features, and preparing the dataset for machine learning.

```
import pandas as pd
import numpy as np
from sklearn.preprocessing import StandardScaler

# Load the Titanic dataset
df = pd.read_csv('train.csv')

# Handle missing values
df['Age'].fillna(df['Age'].median(), inplace=True)
df['Embarked'].fillna(df['Embarked'].mode()[0],
    inplace=True)
```

```
12
13   # Standardize the 'Fare' and 'Age' columns using
14   # NumPy
15   scaler = StandardScaler()
16   df[['Fare', 'Age']] = scaler.fit_transform
17       (df[['Fare', 'Age']])
18
19   # Convert categorical variables into numerical
20   # ones
21   df['Sex'] = df['Sex'].map({'male': 0, 'female':
22       1})
23
24   # Drop irrelevant columns
25   df = df.drop(columns=['Name', 'Ticket', 'Cabin'])
26
27   # Display the cleaned dataset
28   print(df.head())
```

Output

	PassengerId	Survived	Pclass	Sex	Age	SibSp	Parch	Fare	Embarked
0	1	0	3	0	-0.565736	1	0	-0.502445	S
1	2	1	1	1	0.663861	1	0	0.786845	C
2	3	1	3	1	-0.258337	0	0	-0.488854	S
3	4	1	1	1	0.433312	1	0	0.420730	S
4	5	0	3	0	0.433312	0	0	-0.486337	S

Figure 9. Final Cleaned DataSet

Note: Although Numpy is not explicitly used in this example, it is used under the hood by the StandardScale function inside of sklearn.

This example illustrates how Pandas and NumPy can be used together to clean, preprocess, and prepare your dataset. With Pandas handling the structural aspects of the dataset and NumPy handling the numerical transformations, this combination is powerful for building machine learning-ready datasets.

Conclusion

Pandas and **NumPy** are essential libraries for data cleaning and preprocessing. While Pandas allows you to efficiently manipulate and analyze your dataset, NumPy provides the numerical backbone to perform complex mathematical transformations. Together, they form a robust toolkit that simplifies the entire data preprocessing pipeline, from handling missing values to normalizing and standardizing numerical data.

Now that we've explored the power of **Pandas** and **NumPy** for data cleaning and preprocessing, it's time to put everything we've learned into practice. In this hands-on activity, we will apply the techniques discussed throughout the previous chapters to clean and preprocess the **Titanic dataset** from Kaggle. By the end of this activity, you'll have a fully prepared dataset, ready for analysis or to be used in a machine learning model.

Let's walk step-by-step through cleaning the Titanic dataset, handling missing values, removing duplicates, scaling features, and encoding categorical variables using Pandas and NumPy.

Hands-on Activity: Cleaning the Titanic Dataset from Kaggle

In this hands-on activity, we'll clean and preprocess the Titanic dataset using **Pandas** and **NumPy**. Follow along with the code, applying the concepts learned in this book, and by the end, you'll have a clean dataset suitable for building a machine learning model.

Step 1: Load the Dataset

First, ensure you have the Titanic dataset downloaded from Kaggle. If you haven't already, you can download the dataset from the Titanic competition page here[1].

Once downloaded, load the dataset into a Pandas DataFrame.

```
1   import pandas as pd
2
3   # Load the Titanic dataset
4   df = pd.read_csv('train.csv')
5
6   # Display the first few rows of the dataset
7   print(df.head())
```

This will load the dataset and display the first few rows, helping you understand the structure of the data.

Step 2: Handling Missing Values

Next, we'll handle the missing values in the dataset. The columns **Age**, **Cabin**, and **Embarked** contain missing values, so we'll deal with them using various strategies:

- For **Age**, we'll fill missing values with the median age.

---

[1]https://www.kaggle.com/c/titanic

- For **Cabin**, since a large portion of the data is missing, we'll drop this column.
- For **Embarked**, we'll fill missing values with the mode (the most frequent value).

```python
# Check for missing values
print(df.isnull().sum())

# Fill missing 'Age' values with the median
df['Age'].fillna(df['Age'].median(), inplace=True)

# Fill missing 'Embarked' values with the mode
df['Embarked'].fillna(df['Embarked'].mode()[0],
    inplace=True)

# Drop the 'Cabin' column due to excessive
# missing data
df.drop(columns=['Cabin'], inplace=True)

# Verify that missing values have been handled
print(df.isnull().sum())
```

Step 3: Removing Duplicates

Although the Titanic dataset is fairly clean, it's always a good practice to check for duplicate rows.

```
1  # Check for duplicate rows
2  print(f"Number of duplicate rows: {df.duplicated(
3      ).sum()}")
4
5  # If duplicates exist, remove them
6  df.drop_duplicates(inplace=True)
7
8  # Verify duplicates are removed
9  print(f"Number of duplicate rows after removal:
10     {df.duplicated().sum()}")
```

This ensures that no duplicate data distorts your analysis or model.

Step 4: Feature Scaling (Normalization)

Next, we'll normalize the **Fare** and **Age** columns to bring them to the same scale. Since machine learning models often perform better with normalized data, this step is important.

```
1  from sklearn.preprocessing import MinMaxScaler
2
3  # Initialize the MinMaxScaler
4  scaler = MinMaxScaler()
5
6  # Normalize the 'Fare' and 'Age' columns
7  df[['Fare', 'Age']] = scaler.fit_transform
8      (df[['Fare', 'Age']])
9
10 # Display the first few rows of the normalized
11 # data
12 print(df[['Fare', 'Age']].head())
```

At this point, both **Fare** and **Age** have been scaled to values between 0 and 1.

Step 5: Encoding Categorical Variables

Many machine learning algorithms require that all data be numerical. In the Titanic dataset, the **Sex** and **Embarked** columns are categorical, so we'll convert them into numerical format using **label encoding**.

```
1   # Convert 'Sex' to numerical values (0 = male, 1
2   # = female)
3   df['Sex'] = df['Sex'].map({'male': 0, 'female':
4       1})
5
6   # Convert 'Embarked' to numerical values using
7   # one-hot encoding
8   df = pd.get_dummies(df, columns=['Embarked'],
9       drop_first=True)
10
11  # Display the transformed dataset
12  print(df.head())
```

- **Sex** is now binary (0 for male, 1 for female).
- **Embarked** has been one-hot encoded, resulting in new columns like **Embarked_Q** and **Embarked_S**.

Step 6: Dropping Irrelevant Columns

Certain columns like **Name**, **Ticket**, and **PassengerId** are not useful for machine learning models as they don't provide meaningful information for predicting survival. We can drop these columns.

```
1  # Drop irrelevant columns
2  df.drop(columns=['Name', 'Ticket', 'PassengerId']
3      , inplace=True)
4
5  # Display the final cleaned dataset
6  print(df.head())
```

After this step, your dataset is much leaner, containing only the most relevant features for building a machine learning model.

Step 7: Final Check

Before proceeding to modeling, it's a good idea to check the overall status of the dataset one more time.

```
1  # Check for any remaining missing values
2  print(df.isnull().sum())
3
4  # Display basic statistics to ensure everything
5  # is in order
6  print(df.describe())
```

Conclusion

By following these steps, you've successfully cleaned and prepro-cessed the Titanic dataset from Kaggle. You've handled missing values, removed duplicates, scaled numerical features, encoded categorical variables, and dropped irrelevant columns. The dataset is now ready for building machine learning models, and you've practiced applying essential data cleaning techniques using **Pandas** and **NumPy**.

In the next chapter, we'll explore **Data Visualization**, an essential tool for transforming complex data into clear and meaningful visuals. Through charts, graphs, and other visual formats, we can reveal patterns, trends, and insights that are difficult to see in raw numbers. You'll learn about different types of visualizations and how to choose the right one to communicate your data effectively, making it easier to interpret and share key findings with others.

Here are 10 questions to reinforce your understanding of the content in Chapter 3 on "Data Cleaning and Preprocessing," along with a few Python challenges:

Reflective Questions

1. **Identifying Missing Values:**

   - What Python function is used to identify missing values in a DataFrame?

2. **Types of Missing Data:**

   - Explain the difference between Missing Completely at Random (MCAR) and Missing Not at Random (MNAR).

3. **Impact of Missing Values:**

   - How can missing values affect the performance of a machine learning model?

4. **Deletion of Missing Values:**

   - Under what circumstances might you decide to remove rows or columns with missing values instead of imputing them?

5. **Imputation Techniques:**

- What are some common methods for imputing missing numerical data? Describe one.

6. **Advanced Imputation:**

    - Why might someone use model-based imputation instead of simpler methods like mean or median imputation?

7. **Benefits and Limitations:**

    - What are the advantages and potential drawbacks of removing rows with missing values in a dataset?

8. **Practical Application:**

    - If a dataset has a column with 90% missing values, what might be a reasonable handling strategy?

9. **Effect on Analysis:**

    - How can the handling of missing values affect the conclusions you draw from your data analysis?

10. **Dataset Integrity:**

    - After handling missing values, what steps would you take to ensure the integrity of your dataset?

Python Challenges

1. **Identifying Missing Values:**
2. **Handling Missing Values by Imputation:**

- Create a function in Python that replaces all missing values in a specified column of a DataFrame with the median of that column.

3. **Visualization After Imputation**:

- Use matplotlib to visualize the distribution of a feature before and after median imputation.

Chapter 4: Introduction to Data Visualization

Why Visualization Matters

In today's data-driven world, the ability to make sense of vast amounts of data is crucial. However, raw data—whether in spreadsheets or databases—can be difficult to interpret or analyze, especially when dealing with complex datasets. This is where **data visualization** comes in. By transforming data into graphical representations, you can quickly understand trends, patterns, and relationships that might otherwise go unnoticed. In this section, we'll explore why data visualization is such a critical skill in data analytics, and how it enhances decision-making and communication.

1. Simplifying Complex Information

Data visualization allows you to take large, complex datasets and distill them into **easily interpretable visual formats** such as charts, graphs, or maps. Imagine trying to make sense of thousands of rows of raw data in a spreadsheet—this would be overwhelming and time-consuming. Visualization simplifies the data by revealing key insights visually, making it far more accessible and manageable.

- **Example**: A bar chart displaying total sales per region instantly reveals which regions are performing better, something that would take much longer to understand from a table of raw numbers.

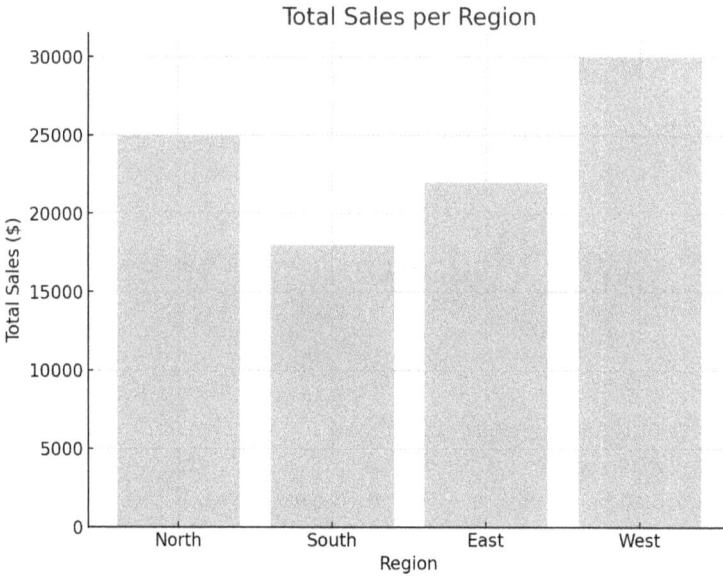

Total Sales per Region

Figure 10. Bar Chart Showing Sales Per Region

2. Identifying Trends and Patterns

One of the key benefits of data visualization is the ability to **detect trends** and **patterns**. Patterns that might remain hidden in raw data can become immediately apparent in a visual representation. Line graphs, for instance, make it easy to track how a particular variable changes over time, enabling users to spot increasing or decreasing trends.

- **Example**: A line graph showing monthly sales over the course of a year can easily highlight periods of growth or decline, helping to identify seasonal patterns or shifts in customer behavior.

Figure 11. Line Graph of Monthly Sales

3. Enhancing Decision-Making

Visualizations don't just present data—they **enhance decision-making** by making it easier to interpret key metrics. By quickly identifying trends, correlations, or outliers, stakeholders can make more informed decisions faster. Whether in business, healthcare, finance, or other fields, data visualization can lead to better outcomes by providing a clearer picture of the current situation.

- **Example**: A company deciding where to invest resources can use a pie chart to visualize which departments consume the most budget and adjust their strategy accordingly.

Department Budget Allocation

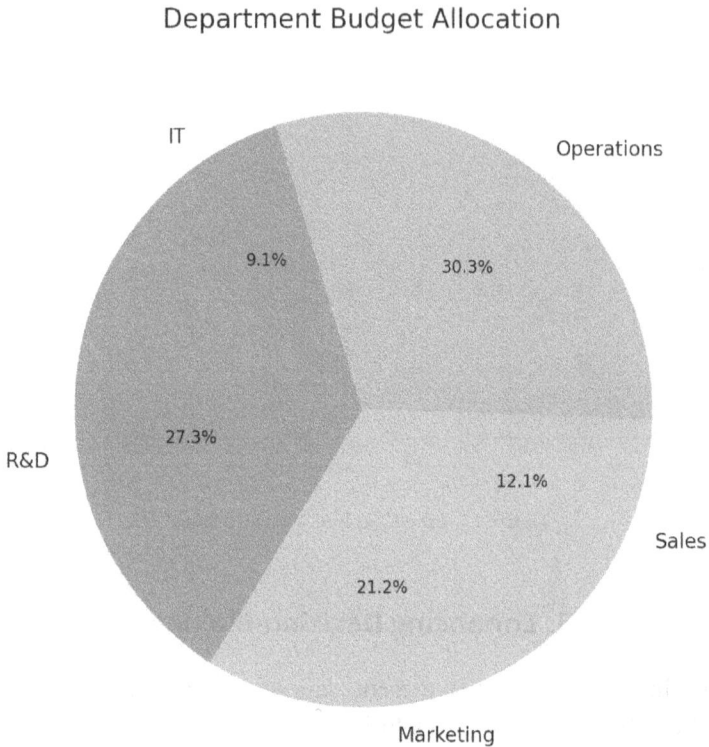

Figure 12. Pie Chart of Department Budgets

4. Communicating Data Effectively

A well-crafted visualization is a powerful communication tool. By turning data into visual stories, you can convey complex information to others more effectively than through raw data alone. Visualizations can make presentations more engaging and help audiences grasp concepts at a glance, especially when sharing insights with non-technical stakeholders.

- **Example**: In a boardroom meeting, using a bar chart to

compare yearly profits across different product lines can make your point much more clearly than presenting a list of numbers.

Yearly Profits Across Product Lines

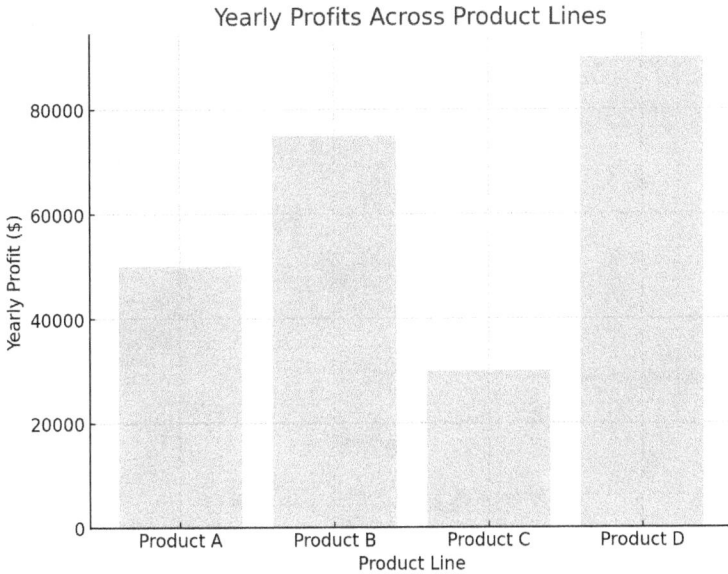

Figure 13. Comparing yearly Profits across Product Lines

5. Finding Outliers and Anomalies

Visualizations are also excellent for spotting **outliers** or **anomalies** in the data, which are often important to address. Scatter plots and box plots, for instance, can visually display where data points deviate significantly from the norm, helping to identify errors, unusual trends, or areas requiring further investigation.

- **Example**: A scatter plot of student test scores versus study hours might reveal a few outliers—students who performed significantly better or worse than expected given their study time.

Test Scores vs Study Hours (with Outliers)

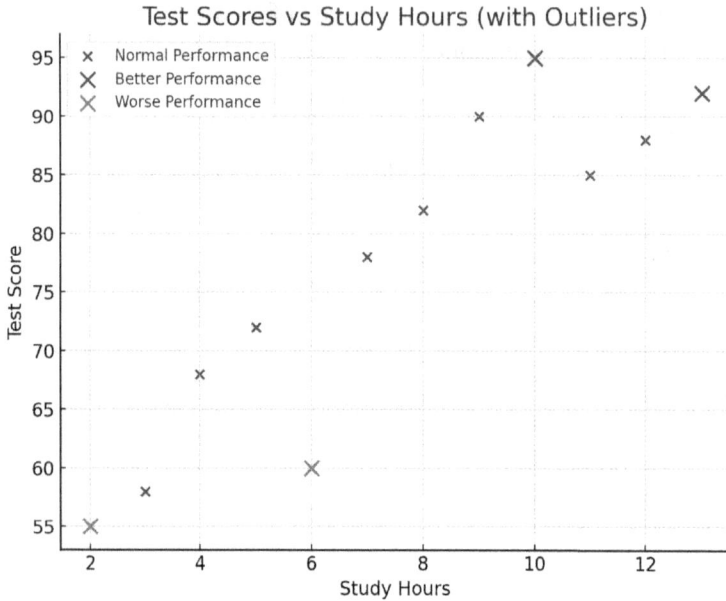

Figure 14. Scatter Plot with Outliers

6. Encouraging Exploration and Interaction

Modern data visualization tools allow users to interact with the data in ways that go beyond static images. **Interactive visualizations** enable users to zoom in on specific data points, filter the data, or drill down into specific categories for deeper analysis. This empowers users to explore the data on their own, facilitating a better understanding of the underlying patterns.

- **Example**: A dashboard that allows users to filter sales data by region, time frame, and product category can help managers gain insights into what's driving sales in specific markets.

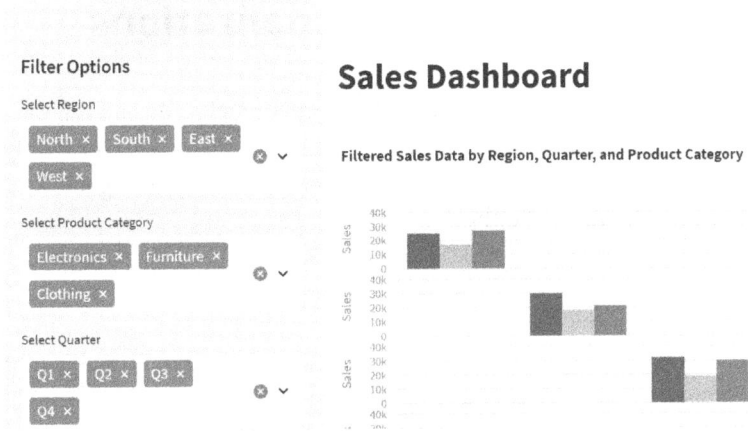

Figure 15. Dashboard Filtering On Sales

Conclusion

Data visualization is much more than just creating pretty charts—it's an essential tool for simplifying complex data, revealing insights, and improving decision-making. Whether you're a data analyst, a business leader, or a researcher, the ability to visualize data effectively is crucial for understanding and communicating information. As we move through this unit, you'll learn how to create meaningful visualizations using various tools and techniques to maximize the impact of your data.

In the next section, we'll introduce different types of visualizations, including bar charts, line graphs, histograms, and scatter plots, and discuss when to use each one.

Types of Data Visualizations

Choosing the right type of visualization is crucial for making data insights clear and accessible. Let's explore the most commonly used data visualizations in analytics, each with Python-generated examples to illustrate their use cases.

1. Bar Charts

A **bar chart** displays data using rectangular bars. It's ideal for comparing different categories, such as sales by product line.

- **When to use**: Use bar charts to compare quantities across categories.
- **Example**: Comparing the total sales of different products.

```
import matplotlib.pyplot as plt

# Bar Chart (Product Sales)
categories = ['Product A', 'Product B',
    'Product C', 'Product D']
values = [50000, 75000, 30000, 90000]

plt.bar(categories, values, color='lightblue')
plt.title('Total Sales per Product')
plt.xlabel('Products')
plt.ylabel('Total Sales ($)')
plt.show()
```

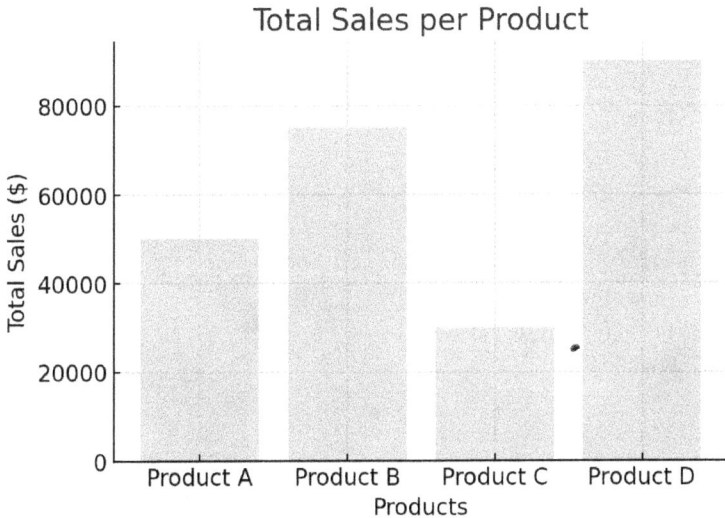

Figure 16. Bar Chart of Product Sales

2. Histograms

A **histogram** shows the distribution of a continuous variable by grouping data into bins. It's useful for understanding how values are spread.

- **When to use**: Use histograms to display the distribution of a continuous variable.
- **Example**: Showing the age distribution of customers.

```
1   import matplotlib.pyplot as plt
2   # Histogram (Age Distribution)
3   ages = [22, 35, 28, 45, 50, 23, 27, 42, 36, 29,
4       21, 33, 40, 24, 41]
5
6   plt.hist(ages, bins=5, color='lightgreen',
7       edgecolor='black')
8   plt.title('Customer Age Distribution')
9   plt.xlabel('Age')
10  plt.ylabel('Frequency')
11  plt.show()
```

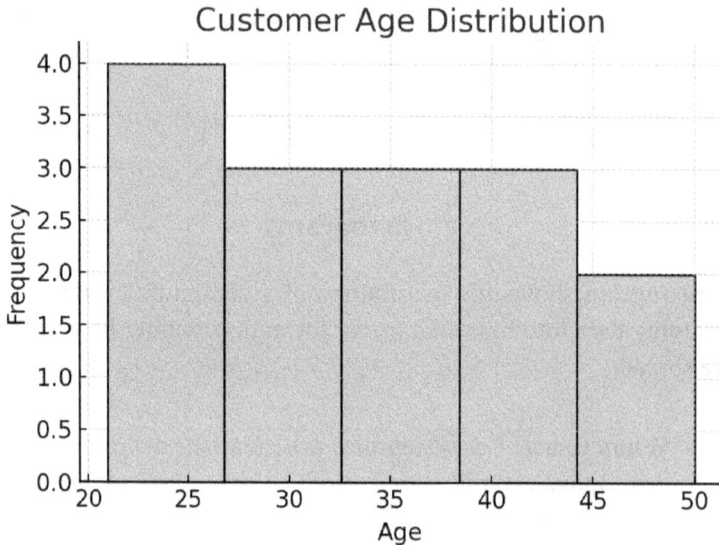

Figure 17. Histogram of Age Distribution

3. Line Graphs

A **line graph** is great for tracking changes over time. It's particularly effective for time-series data like monthly sales or stock prices.

- **When to use**: Use line graphs to visualize trends over time.
- **Example**: Tracking monthly sales over a year.

```
1   import matplotlib.pyplot as plt
2   # Line Chart (Monthly Sales)
3   months = ['Jan', 'Feb', 'Mar', 'Apr', 'May',
4       'Jun', 'Jul', 'Aug', 'Sep', 'Oct', 'Nov',
5       'Dec']
6   sales = [10000, 12000, 15000, 13000, 14000, 16000
7       , 18000, 17500, 20000, 21000, 22000, 25000]
8
9   plt.plot(months, sales, marker='o',
10      color='purple')
11  plt.title('Monthly Sales Over the Year')
12  plt.xlabel('Month')
13  plt.ylabel('Sales ($)')
14  plt.show()
```

Figure 18. Line Chart of Sales

4. Scatter Plots

A **scatter plot** is ideal for examining the relationship between two continuous variables. It's often used to detect correlations.

- **When to use**: Use scatter plots to show relationships or correlations between two variables.
- **Example**: Visualizing the relationship between study hours and exam scores.

```
1   import matplotlib.pyplot as plt
2   # Scatter Plot (Study Hours vs Exam Scores)
3   study_hours = [2, 3, 4, 5, 6, 7, 8, 9, 10, 11, 12
4       , 13]
5   exam_scores = [55, 60, 65, 70, 75, 80, 82, 85,
6                  90, 95, 92, 97]
7
8   plt.scatter(study_hours, exam_scores,
9       color='darkblue')
10  plt.title('Study Hours vs Exam Scores')
11  plt.xlabel('Study Hours')
12  plt.ylabel('Exam Scores')
13  plt.show()
```

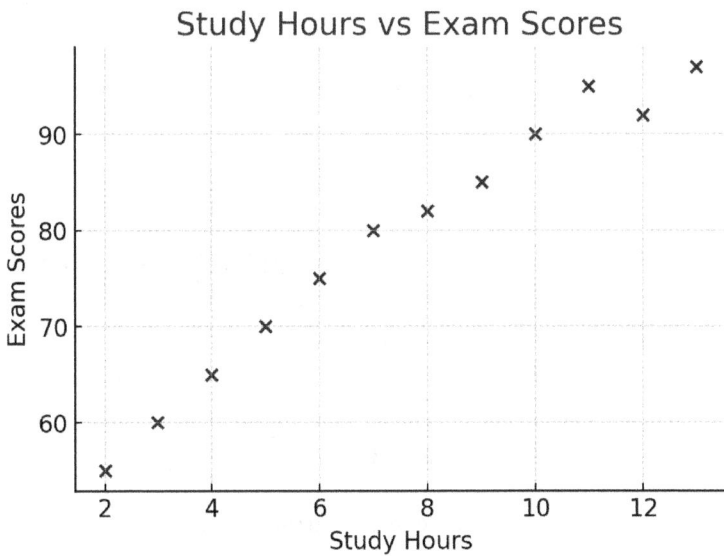

Figure 19. Scatter Plot

Box Plot

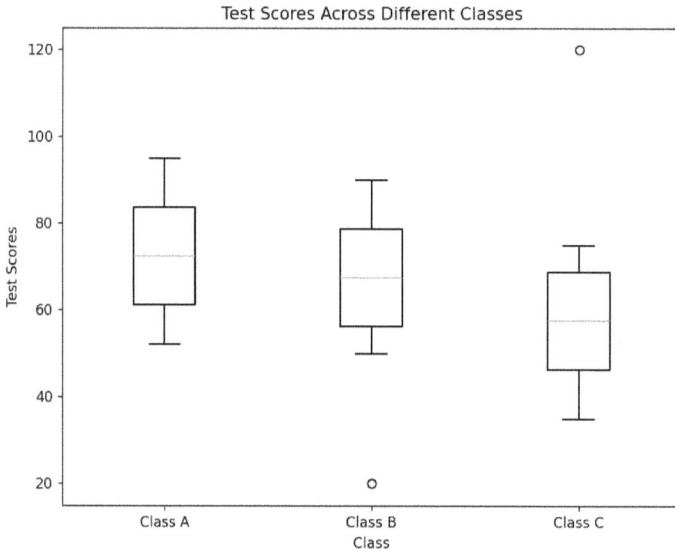

Figure 20. Boxplot of Test Scores

A **box plot** (or whisker plot) is used to display the distribution of a dataset by showing the dataset's quartiles (the 25th, 50th, and 75th percentiles) and highlighting any **outliers**. It's particularly useful for comparing distributions between groups or categories.

In the example above, the box plot compares the **test scores** across three different classes (Class A, Class B, and Class C).

- The **box** represents the interquartile range (IQR), which contains the middle 50% of the data.
- The **line** inside the box shows the **median** (50th percentile) of the data.
- The **whiskers** extend to the minimum and maximum data points within 1.5 times the IQR, while any points outside this range are considered **outliers**.

When to Use a Box Plot:

- To understand the spread of your data.
- To compare distributions between different categories or groups.
- To identify **outliers** or extreme values that fall outside the expected range.

Box plots are especially useful when you have multiple datasets or groups to compare, as they provide a quick overview of the range, median, and potential outliers in each dataset.

Here is the full code for generating a **box plot** in Python using `matplotlib`:

```
1   # Import necessary libraries
2   import matplotlib.pyplot as plt
3
4   # Example data for box plot (test scores of
5   # students in different classes)
6   class_a_scores = [52, 55, 60, 65, 70, 75, 80, 85,
7       90, 95]
8   class_b_scores = [20, 50, 55, 60, 65, 70, 75, 80,
9       85, 90]
10  class_c_scores = [35, 40, 45, 50, 55, 60, 65, 70,
11      75, 120]
12
13  # Data to plot
14  data = [class_a_scores, class_b_scores,
15      class_c_scores]
16
17  # Create a box plot
18  plt.figure(figsize=(8, 6))
19  plt.boxplot(data)
20
21  # Add labels and title
```

```
22   plt.title('Test Scores Across Different Classes')
23   plt.xlabel('Class')
24   plt.ylabel('Test Scores')
25   plt.xticks([1, 2, 3], ['Class A', 'Class B',
26       'Class C'])
27
28   # Show plot
29   plt.show()
```

Running this code will generate a box plot comparing the test scores of three different classes. Note that both Class B and Class C have outliers, represented by small circles appearing below and above the whiskers, respectively.

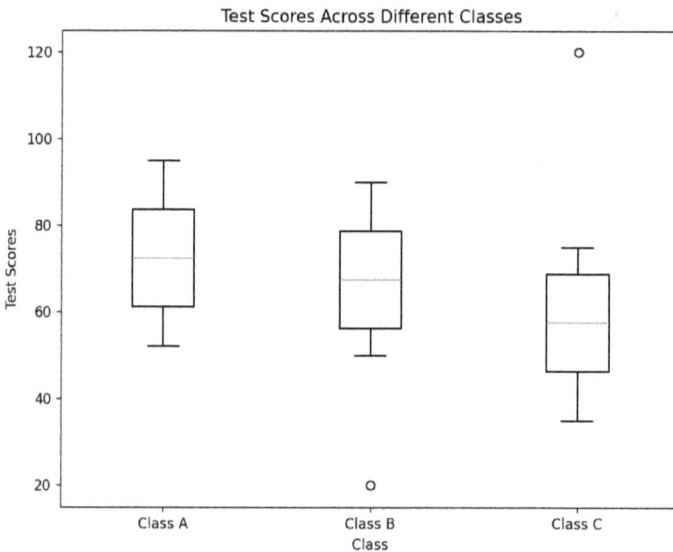

Figure 21. Boxplot of Test Scores

Conclusion

Different types of data visualizations allow you to convey your insights more effectively, depending on the nature of your data. Whether you're comparing categories with a bar chart, tracking trends with a line graph, or exploring relationships with a scatter plot, visualizations play a crucial role in making data understandable and actionable.

Example of Data Storytelling: "The Impact of Marketing Spend on Revenue Growth"

The Story:

Imagine you're presenting the results of a year-long marketing campaign to your company's leadership team. Your goal is to show the direct impact of marketing spend on revenue and highlight key insights that can drive future strategy.

Setting the Context:

"We've been working hard on our marketing strategy this year, investing in various channels to boost our revenue. Today, I'll walk you through how our marketing spend has translated into actual revenue growth and identify which efforts had the most significant impact."

Introducing the Data:

You start by showing a **line graph** that tracks **monthly marketing spend** and **revenue** over the past 12 months.

Figure 22. **Storytelling Revenue**

- The **line graph** visually shows that as marketing spend increased steadily from January to June, revenue followed a similar upward trend, reaching its highest point in July.
- You can highlight a **spike in revenue** in July, corresponding to a particular marketing campaign.

Highlighting Key Insights:

"As you can see from this chart, the more we invested in our marketing efforts, the greater the revenue growth. In particular, the major jump in July was due to the launch of our social media campaign, which boosted our engagement and led to a 25% increase in revenue that month."

At this point, you zoom in on a **scatter plot** that breaks down the **relationship between marketing spend by channel** (social media, email, search ads) and revenue growth.

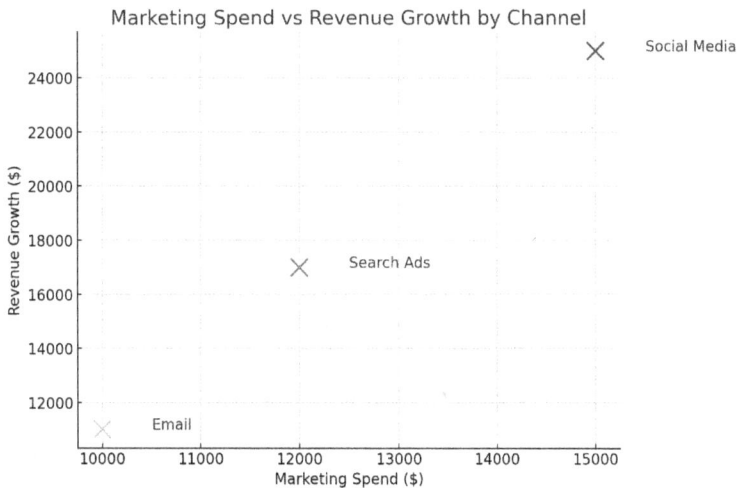

Figure 23. Market Spending

- You point out that **social media campaigns** had the highest return on investment (ROI), while **email campaigns** had a lower-than-expected impact despite significant investment.

Addressing Outliers:

Next, you use a **scatter plot** to highlight an important outlier: "While most of our campaigns followed a predictable pattern—more spending equals more revenue—take a look at this outlier in November. We increased our ad spend significantly, but revenue dipped."

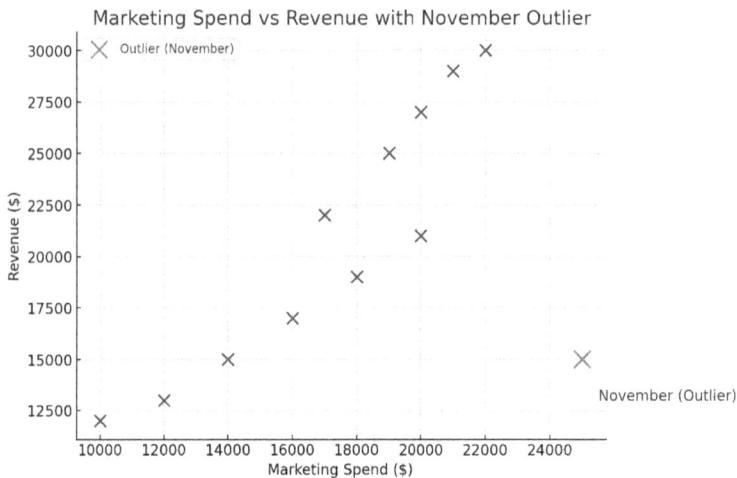

Figure 24. Ad Spend Scatter Plot

You explain the outlier by adding some context: "This dip occurred because our search ads weren't optimized, leading to low conversion rates despite higher spend."

Conclusion and Call to Action:

"Based on these insights, we can see that social media continues to be our strongest channel. I recommend reallocating some of our email marketing budget to social media for the next quarter.

Additionally, we should review our search ad strategy to improve conversions and get a better return on investment."

Key Elements of the Story:

1. **Setting the context**: You start by establishing the goal—showing the impact of marketing spend on revenue growth.
2. **Introducing data with visuals**: Use **line graphs** and **scatter plots** to show trends, outliers, and the relationship between marketing channels and revenue.
3. **Highlighting key insights**: Focus on how specific campaigns contributed to revenue growth, and use the data to explain successes and failures.
4. **Adding human context**: Explain why certain trends occurred and what decisions were made to influence the data.
5. **Call to action**: End the story with a recommendation based on the data insights, motivating the team to take action.

This is an example of how to use **data storytelling** to not only present data but also craft a narrative that guides your audience to understand the key insights and take meaningful action.

Here are some reflective questions to reinforce your understanding of the content in Chapter 4 on "Introduction to Data Visualization," along with a few Python challenges:

Reflective Questions

1. **Visual Simplification**:

   • How does data visualization simplify the understanding of complex datasets?

2. **Trends and Patterns:**

   - Why is it easier to identify trends and patterns with visualizations compared to raw data?

3. **Decision-Making:**

   - How can visualizing data enhance decision-making in a business context?

4. **Effective Communication:**

   - Discuss how a well-crafted visualization can improve communication with non-technical stakeholders.

5. **Outliers Identification:**

   - What role do visualizations play in identifying outliers or anomalies in data?

6. **Interactive Visualizations:**

   - What advantages do interactive visualizations offer over static images in the context of data exploration?

7. **Visualization Tools:**

   - What are some common tools or software that can be used for creating data visualizations?

8. **Choice of Visualization:**

   - Explain how you would decide which type of visualization to use for a given dataset.

9. **Ethical Considerations:**

- What are some ethical considerations to keep in mind when creating data visualizations?

10. **Future Trends:**

- Speculate on how the field of data visualization might evolve in the next five years.

Python Challenges

1. **Creating a Bar Chart:**

- Write a Python script using matplotlib to create a bar chart that compares the average monthly sales data across four different product categories.

```
1   import matplotlib.pyplot as plt
2
3   # Data
4   categories = ['Electronics', 'Clothing', 'Furniture', 'To\
5   ys']
6   sales = [15000, 22000, 13000, 18000]
7
8   # Bar Chart
9   # <add your code here>
```

2. **Generating a Line Graph:**

- Create a Python script to plot a line graph showing the growth in subscriber count over twelve months.

```
1    import matplotlib.pyplot as plt
2
3    # Data
4    months = ['Jan', 'Feb', 'Mar',
5              'Apr', 'May', 'Jun',
6              'Jul', 'Aug', 'Sep',
7              'Oct', 'Nov', 'Dec']
8    subscribers = [120, 150, 180, 210,
9       240, 270, 300, 330, 360,
10      390, 420, 450]
11
12   # Line Graph
13   # <add your code here>
```

3. **Creating a Scatter Plot:**

   - Use Python to create a scatter plot that shows the relationship between advertising spend and revenue.

```
1    import matplotlib.pyplot as plt
2
3    # Data
4    ad_spend = [500, 600, 700, 800, 900, 1000,
5                1100, 1200, 1300, 1400]
6    revenue = [2000, 2100, 2200, 2300,
7                2400, 2500, 2600, 2700, 2800, 2900]
8
9    # Scatter Plot
10   # <add your code here>
```

Chapter 5: Tools for Visualization

Introduction to `Matplotlib` and `Seaborn`

Data visualization plays a crucial role in understanding and interpreting data, and Python offers several powerful libraries to create clear, engaging visual representations. Two of the most widely used libraries for data visualization in Python are **Matplotlib** and **Seaborn**. These libraries provide flexible and easy-to-use tools that allow you to create a wide range of visualizations, from simple plots to complex statistical graphics.

In this section, we'll introduce you to both `Matplotlib` and `Seaborn`, covering their strengths and how to use them effectively for visualizing data in Python.

1. `Matplotlib`: The Foundation of Python Visualization

`Matplotlib` is a versatile and foundational plotting library for Python. It allows users to create a wide variety of visualizations, from line graphs and bar charts to scatter plots and histograms. As one of the earliest and most widely used libraries in the Python ecosystem, `Matplotlib` provides extensive control over the appearance of plots, making it a go-to tool for detailed and customized visualizations.

Key Features of `Matplotlib`**:**

- **Customizability**: `Matplotlib` gives you full control over every aspect of your plots, from color schemes and marker styles to axis labels and legends.
- **Wide Range of Plots**: You can create basic plots like bar charts, scatter plots, and line graphs, as well as more complex visuals like 3D plots and pie charts.
- **Integration**: `Matplotlib` works seamlessly with other Python libraries, such as `NumPy` and `Pandas`, to generate plots directly from data structures like arrays and DataFrames.

Example: Basic Plot using `Matplotlib`

Here's an example of how to create a simple line plot using Matplotlib:

```
1   import matplotlib.pyplot as plt
2
3   # Example data: months and sales
4   months = ['Jan', 'Feb', 'Mar', 'Apr', 'May']
5   sales = [10000, 12000, 15000, 17000, 20000]
6
7   # Create a basic line plot
8   plt.plot(months, sales, marker='o', color='blue')
9
10  # Add title and labels
11  plt.title('Monthly Sales Over Time')
12  plt.xlabel('Month')
13  plt.ylabel('Sales ($)')
14  plt.show()
```

In this example, `Matplotlib` is used to plot monthly sales data, generating a clear and simple line graph. This is just a glimpse of the power and flexibility `Matplotlib` provides.

2. Seaborn: **Simplifying Statistical Visualizations**

While Matplotlib offers flexibility, it can sometimes require more code for complex visualizations. This is where **Seaborn** comes in. Built on top of Matplotlib, Seaborn simplifies the process of creating attractive and informative statistical graphics. It provides a higher-level interface that automatically applies better aesthetics and can handle more complex visualizations with less effort.

Key Features of Seaborn:

- **Built-in Themes**: Seaborn comes with a variety of attractive color palettes and themes that automatically improve the appearance of your plots.
- **Statistical Plots**: Seaborn specializes in visualizing statistical relationships, making it ideal for creating heatmaps, violin plots, and pair plots, among others.
- **Integration with Pandas**: Seaborn works seamlessly with Pandas DataFrames, making it easier to generate complex visualizations from structured data.

Example: Creating a Scatter Plot with Regression Line using Seaborn

Here's an example using Seaborn to visualize the relationship between two variables and add a regression line:

```
1   import seaborn as sns
2   import matplotlib.pyplot as plt
3
4   # Example data
5   # load a csv file of tips data
6   tips = sns.load_dataset('tips')
7
8   # Create a scatter plot with a regression line
9   sns.lmplot(x='total_bill', y='tip', data=tips)
10
11  # Add title
12  plt.title('Total Bill vs. Tip Amount')
13  plt.show()
```

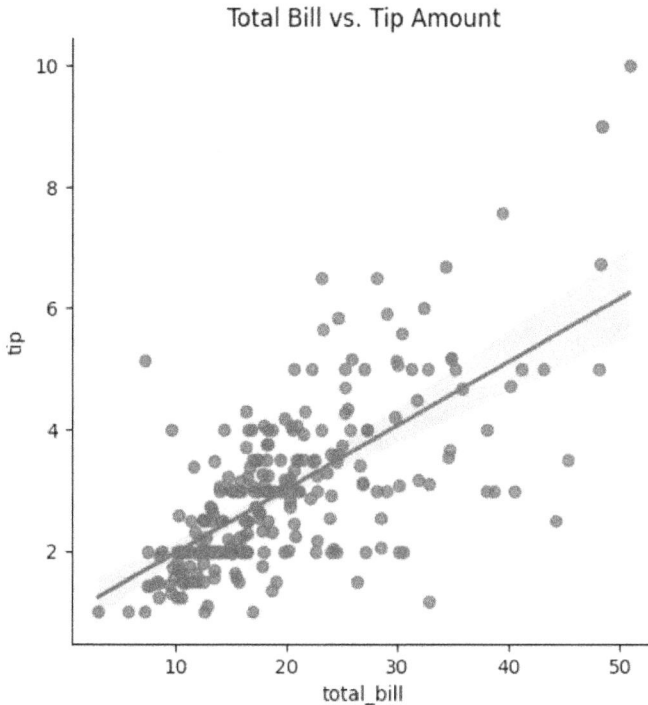

Figure 25. Scatter Plot with Regression

In this example, Seaborn makes it easy to create a scatter plot with a regression line, showcasing the relationship between total bill and tip amount in a restaurant dataset. The use of built-in themes and integration with Pandas allows for cleaner and more concise code.

Here's a snapshot of our **tips dataset**, which contains detailed information about customer transactions at a restaurant. Each row represents a unique transaction, and the columns provide insights like the **total bill**, **tip amount**, and other details such as the **day of the week**, **time of day**, **customer gender**, and whether the customer was a **smoker**.

One of the benefits of using **Seaborn** is that it can read and visualize data directly from this dataset, without requiring additional steps to bring the data into a DataFrame first, making it easy to generate quick, meaningful insights.

Tips Dataset

	total_bill	tip	sex	smoker
1	21.854305348131312	6.77828481538859	Male	No
2	47.78214378844623	1.7572596849559494	Female	Yes
3	37.93972738151323	2.454658426851524	Male	Yes
4	31.93963178886665	9.086987696743714	Male	Yes
5	12.020838819909644	6.457861536936309	Female	Yes
6	12.0197534151291	1.08277346454966	Female	Yes

Figure 26. Tips Data

Choosing Between Matplotlib and Seaborn

Both Matplotlib and Seaborn are essential tools for data visualization in Python, but they serve slightly different purposes:

- Use Matplotlib when you need complete control over your plots and want to build highly customized visualizations.
- Use Seaborn when you need to create clean, statistical plots quickly and easily, especially when working with structured data like Pandas DataFrames.

In practice, `Seaborn` can be used for many high-level tasks, while `Matplotlib` can step in for more detailed customization, making them complementary tools.

In the next section, we will dive deeper into **creating specific types of visualizations using `Matplotlib` and `Seaborn`**, exploring their real-world applications in data analysis.

Visualizing Distributions, Relationships, and Comparisons

When working with data, visualizing its distribution, relationships between variables, and comparisons across categories are essential to uncover patterns and insights. In this section, we will explore how to use **Seaborn** and **Matplotlib** to create visualizations that help you understand your data from these perspectives.

---

1. Visualizing Distributions

Visualizing the distribution of your data allows you to see how values are spread across a range and identify patterns such as skewness, clustering, or the presence of outliers.

Common Plots for Distribution:

- **Histograms:** Useful for showing the distribution of a single continuous variable by dividing it into intervals (bins).

- **KDE (Kernel Density Estimate) Plots**: A smoothed curve that represents the distribution of a continuous variable.
- **Box Plots**: Summarizes the distribution of a variable by showing its quartiles and potential outliers.

Example: Visualizing the distribution of total bills in the tips dataset using a histogram and KDE plot:

```
import seaborn as sns
import matplotlib.pyplot as plt

# Create a histogram with KDE for the
# 'total_bill' column
sns.histplot(tips['total_bill'], kde=True,
    color='blue')

# Add title and labels
plt.title('Distribution of Total Bill Amounts')
plt.xlabel('Total Bill ($)')
plt.ylabel('Frequency')
plt.show()
```

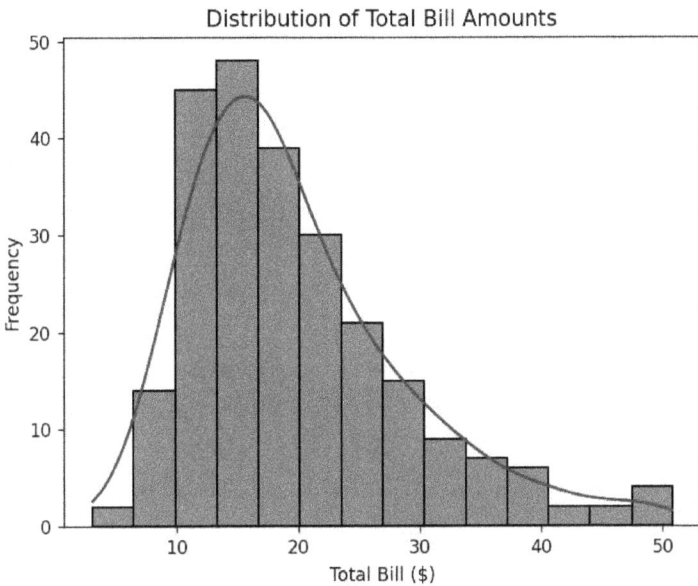

Figure 27. Histogram

This plot shows the spread of **total bill amounts**, allowing you to see whether most bills cluster around a particular range and how the distribution looks overall.

---

2. Visualizing Relationships

To understand the relationship between two variables, scatter plots and regression lines are some of the best tools. These allow you to visualize potential correlations or patterns between variables, such as how the **total bill** might influence the **tip** amount in the tips dataset.

Common Plots for Relationships:

- **Scatter Plots**: Plot individual data points to reveal relationships or correlations between two variables.
- **Line Plots**: Used when you want to plot a continuous relationship over time or between two variables.
- **Pair Plots**: A grid of scatter plots for every pair of variables in the dataset.

Example: Visualizing the relationship between meal size and tip:

```
1   # Create a line plot between size and tip
2   sns.lineplot(data=tips, x='size', y='tip')
3
4   # Add title
5   plt.title('Relationship Between Size and Tip')
6   plt.show()
```

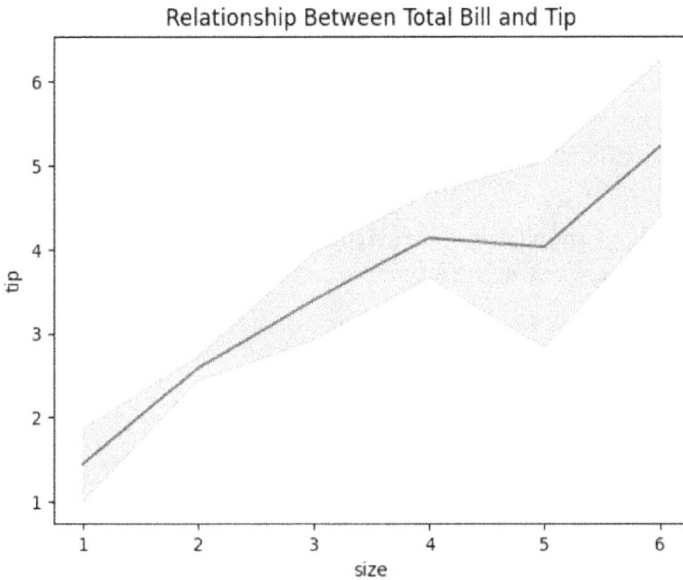

Figure 28. Line Plot for Size and Tip

This line plot shows how **size** and **tip** are related, helping you see whether larger sizes tend to result in larger tips.

---

3. Visualizing Comparisons

When comparing data across different categories, bar charts, box plots, and violin plots can help you see how different groups perform against each other.

Common Plots for Comparisons:

- **Bar Plots**: Compare the mean (or other aggregate measure) across categories.

- **Box Plots**: Compare the distribution of a variable across multiple categories.
- **Violin Plots**: Combines the features of a box plot and a KDE plot, showing both the distribution and summary statistics of a variable across categories.

Example: Comparing the distribution of tips across different days of the week using a box plot:

```
# Create a box plot to compare tips across
# different days
sns.boxplot(x='day', y='tip', data=tips)

# Add title
plt.title('Tip Amounts by Day of the Week')
plt.xlabel('Day of the Week')
plt.ylabel('Tip Amount ($)')
plt.show()
```

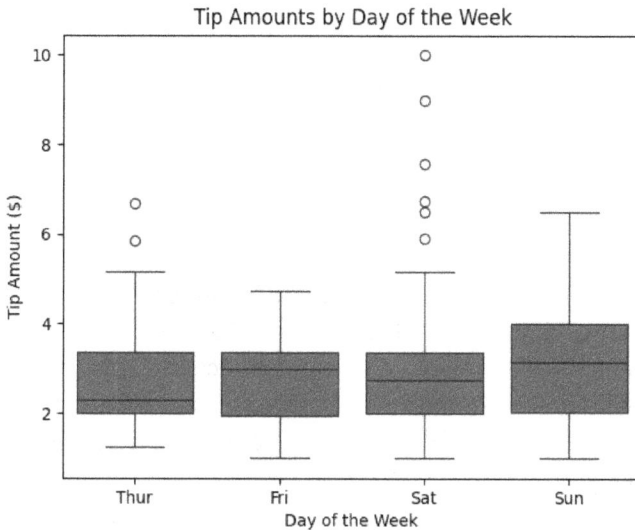

Figure 29. Box Plot for Tips

The **box plot** shows **tip amounts by day of the week**. Here's the analysis based on the visual representation:

1. Median Tip Amount:

- The **median** (the line inside each box) indicates that the median tip is fairly consistent across **Thursday, Friday**, and **Saturday**, with all days hovering between 2.5 and 3 dollars.
- **Sunday** has a slightly higher median, suggesting that customers tend to leave slightly larger tips on Sundays compared to the other days.

2. Spread of Tips (IQR):

- The **Interquartile Range (IQR)**, represented by the height of the box, shows the middle 50% of tip values for each day.

- **Friday** has the smallest IQR, indicating that tipping behavior is more consistent on that day.
- **Sunday** has the widest IQR, showing more variability in tipping behavior on that day, with a wider range of tip amounts.

3. Outliers:

- **Saturday** has the most **outliers**, represented by the dots above the whiskers. This suggests that there are some customers on Saturdays who tip significantly more than the majority of customers.
- **Thursday** and **Friday** also have a few outliers, but not as many as Saturday.
- **Sunday** does not show many extreme outliers, but the box and whiskers suggest a wider overall range of tips.

4. Whiskers:

- The whiskers represent the overall range of tips, excluding outliers.

  - **Friday** has the shortest whiskers, meaning the range of tip amounts on Friday is narrower and more consistent.
  - **Sunday** and **Thursday** have longer whiskers, showing a wider range of tip amounts, even before accounting for outliers.

Key Observations:

- **Sunday** has the highest median and the largest spread of tips, indicating that while tips may vary more, customers tend to tip a bit more generously on Sundays.

- **Saturday** shows the most **outliers**, with some customers tipping significantly higher than the majority.
- **Friday** is the most consistent day, with fewer outliers and a narrower spread of tips.
- **Thursday** has moderate consistency, with a few outliers but a relatively typical tip distribution.

This analysis could suggest that **Saturdays** attract more customers who occasionally leave **large tips**, while **Sundays** see generally **higher** but **more variable tipping**. **Fridays** have the most **predictable tipping patterns**, likely showing more regular, consistent tipping behavior.

Violin Plot

This code generates a violin plot that shows how tip amounts are distributed across different days of the week. The **quartiles** are included to help highlight the spread of the data.

```python
# Creating a violin plot to visualize the
# distribution of tips by day of the week
sns.violinplot(x='day', y='tip', data=tips,
    inner='quartile', palette='coolwarm')

# Add title and labels
plt.title(
    'Distribution of Tip Amounts by Day of \
    the Week')
plt.xlabel('Day of the Week')
plt.ylabel('Tip Amount ($)')

# Adjust layout
plt.tight_layout()

# Display the plot
plt.show()
```

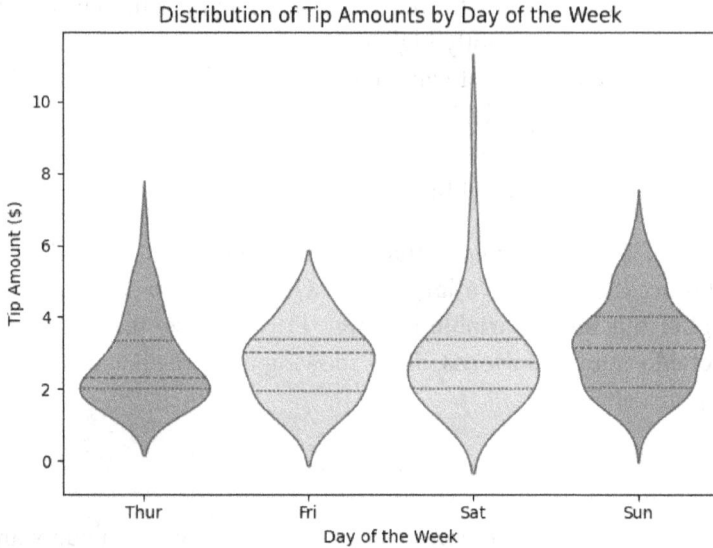

Figure 30. Violin Plot showing Distributions

Here is a violin plot showing the distribution of tip amounts by day of the week. The plot displays the distribution of tips for each day, with the width of the violin indicating the density of the data. The inner lines represent the quartiles, helping you understand the spread and central tendency of tips on different days. This plot provides both the distribution and statistical insights, making it useful for comparing tipping patterns across the week.

Analysis of the Violin Plot: Distribution of Tip Amounts by Day of the Week

The **violin plot** provides a more detailed visualization of the distribution of **tip amounts** across different days of the week. It combines aspects of both a **box plot** and a **kernel density plot**, giving us a clear picture of both the distribution and frequency of tip amounts.

1. Overall Shape of the Distributions:

- The width of each "violin" shows the **density** of tip amounts for that particular day. Wider sections of the violin represent more frequent tip values, while narrower sections indicate less frequent tip values.
- **Saturday** has the tallest and widest distribution, showing a broader range of tip values compared to the other days.

2. Median Tip Amount (Dashed Line):

- The **dashed line** inside each violin represents the **median tip amount** for that day. As in the box plot, the median is relatively consistent across the week, with **Sundays** appearing to have a slightly higher median than the other days.
- **Saturday** has a slightly lower median compared to **Sunday**, but still within the same general range.

3. Distribution of Tips:

- **Thursday** and **Friday**: Both days show a relatively **narrow distribution**, meaning tips are clustered around the median. This suggests **more consistent tipping behavior** on these days, with fewer extreme variations.
- **Saturday**: The violin plot for Saturday has a **wider and taller distribution**, indicating a wider **range of tipping behavior**. The sharp peak at the bottom suggests a cluster of low tips, but the long tail extending upwards shows that a small number of customers leave much higher tips, likely explaining the outliers seen in the previous box plot.
- **Sunday**: Sunday's violin plot shows a relatively **even distribution** of tips, with fewer low tips and a moderate range of higher tips.

4. Skewness of Distribution:

- **Saturday** is skewed towards **higher tip amounts**, with a visible tail extending upward, indicating that while many customers leave moderate tips, there is a significant group leaving very high tips.
- **Thursday** and **Friday** have a more **symmetric distribution**, meaning that tips are generally centered around the median, without a significant skew toward very high or very low values.

5. Frequency of Tipping Behavior:

- The width of the violins at different levels represents the **frequency** of those tip amounts.

  - For example, **Thursday** has a noticeable bulge near the **$2.5 to $3 range**, meaning that many customers tend to leave tips in this range.
  - **Saturday** shows a **wider spread** in tip amounts, suggesting more variability in tipping behavior, with some customers leaving lower tips while others leave much higher tips.

Key Observations:

- **Saturday** has the **most variable tipping behavior**, with a wider range of both low and high tips compared to the other days.
- **Sunday** shows relatively higher tips, with a distribution more skewed toward moderate to high tips.
- **Thursday** and **Friday** show more **consistent tipping behavior**, with tips generally clustered around the median and fewer extreme values.

In summary, **Saturdays** attract a wider variety of tip amounts, suggesting that both very generous and lower tips are common. Meanwhile, **Thursdays** and **Fridays** are more predictable in terms of tipping behavior, with customers tending to tip within a narrower range. **Sundays** seem to attract somewhat higher tips, but with a more balanced distribution.

Conclusion

Visualizing distributions, relationships, and comparisons helps you uncover hidden patterns in your data. By using the right visualizations, such as histograms for distributions, scatter plots for relationships, and box plots for comparisons, you can gain deeper insights and make data-driven decisions with confidence. In the next section, we'll dive deeper into how to customize these visualizations to better fit your analysis and reporting needs.

——-

Got it! Let's focus strictly on the **visualization** of the **Iris dataset** for this section.

Hands-on Activity: Visualizing the Iris Dataset from Kaggle

In this activity, we will use Python to visualize different features of the **Iris dataset**. This dataset includes measurements of three species of Iris flowers: **Setosa, Versicolor**, and **Virginica**. The goal of this activity is to visualize the relationships between the four key features—**sepal length, sepal width, petal length**, and **petal width**—and how these features differ across species.

Step 1: Load the Iris Dataset

We will start by loading the Iris dataset. You can use the built-in version available in the **Seaborn** library, which includes the dataset:

```python
import seaborn as sns
import pandas as pd

# Load the Iris dataset
iris_df = sns.load_dataset('iris')

# Display the first few rows of the dataset
iris_df.head()
```

This will load the Iris dataset, showing the following columns:

- **sepal_length**: The length of the sepal in centimeters.
- **sepal_width**: The width of the sepal in centimeters.
- **petal_length**: The length of the petal in centimeters.
- **petal_width**: The width of the petal in centimeters.
- **species**: The species of Iris flower (Setosa, Versicolor, or Virginica).

Step 2: Visualizing Distributions of Features

Let's start by visualizing the distributions of the key features (sepal and petal measurements) to understand the range and frequency of each variable.

2.1: Histograms for Sepal Length and Petal Length

```
1   import matplotlib.pyplot as plt
2
3   # Create histograms for sepal length and petal
4   # length
5   plt.figure(figsize=(10, 5))
6
7   # Sepal Length
8   plt.subplot(1, 2, 1)
9   sns.histplot(
10      iris_df['sepal_length'],
11      kde=True,
12      color='blue')
13  plt.title('Distribution of Sepal Length')
14
15  # Petal Length
16  plt.subplot(1, 2, 2)
17  sns.histplot(
18      iris_df['petal_length'],
19      kde=True,
20      color='green')
21  plt.title('Distribution of Petal Length')
22
23  plt.tight_layout()
24  plt.show()
```

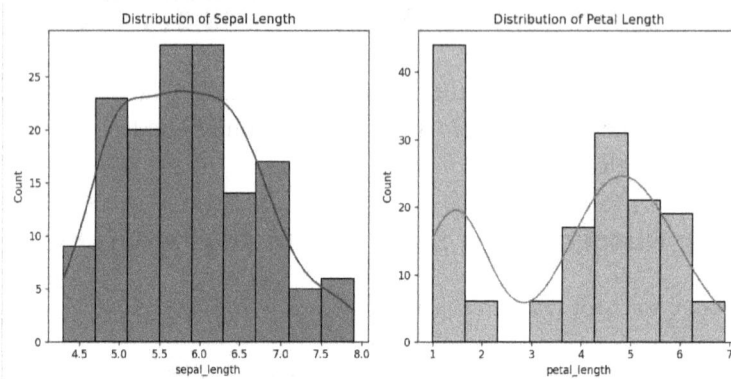

Figure 31. **Distributions of Sepals and Petals**

These histograms show the distribution of **sepal length** and **petal length** across all samples. The **KDE (Kernel Density Estimation)** curves help visualize the probability distribution of the data.

Step 3: Scatter Plots for Feature Relationships

Scatter plots are ideal for visualizing relationships between pairs of numerical features. Let's create scatter plots to see how features like **petal length** and **petal width** vary between species.

3.1: Petal Length vs. Petal Width by Species

```
1    # Scatter plot of petal length vs petal width,
2    # colored by species
3    plt.figure(figsize=(8, 6))
4    sns.scatterplot(
5        x='petal_length',
6        y='petal_width',
7        hue='species',
8        data=iris_df)
9    plt.title('Petal Length vs. Petal Width by \
10       Species')
11   plt.xlabel('Petal Length (cm)')
12   plt.ylabel('Petal Width (cm)')
13   plt.show()
```

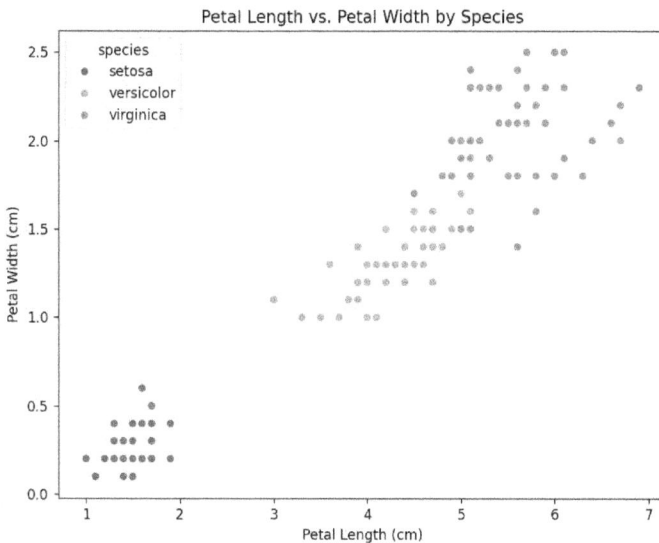

Figure 32. Petal Length and Width Scatter Plots

This scatter plot shows the relationship between **petal length** and **petal width** for each species. By coloring the points based on the species, we can easily see how these two features differ across Setosa, Versicolor, and Virginica.

Step 4: Pair Plot to Visualize All Relationships

A **pair plot** is a useful tool to visualize the pairwise relationships between all features in the dataset. It provides scatter plots for every pair of features and histograms for the distribution of each feature.

```
# Pair plot to visualize relationships between
# all features
sns.pairplot(iris_df, hue='species')
plt.suptitle(
  'Pairwise Relationships in the Iris Dataset',
   y=1.02)
plt.show()
```

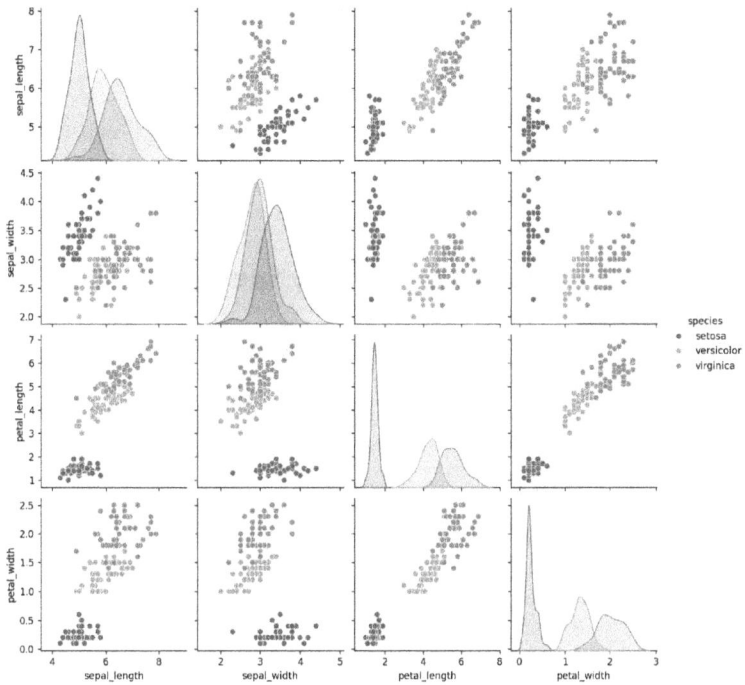

Figure 33. Pair Plot Comparison

The pair plot shows:

- Scatter plots for each pair of features (e.g., **sepal length** vs. **sepal width**).
- Histograms along the diagonal, representing the distribution of each feature.
- Different colors for each species, making it easier to see how species are separated based on the features.

Step 5: Box Plots to Compare Feature Distribution Across Species

Box plots are ideal for comparing the distribution of numerical features across different categories, such as the three species in the Iris dataset.

5.1: Sepal Width by Species

```
1  # Box plot of sepal width by species
2  plt.figure(figsize=(8, 6))
3  sns.boxplot(x='species', y='sepal_width',
4      data=iris_df)
5  plt.title('Sepal Width by Species')
6  plt.xlabel('Species')
7  plt.ylabel('Sepal Width (cm)')
8  plt.show()
```

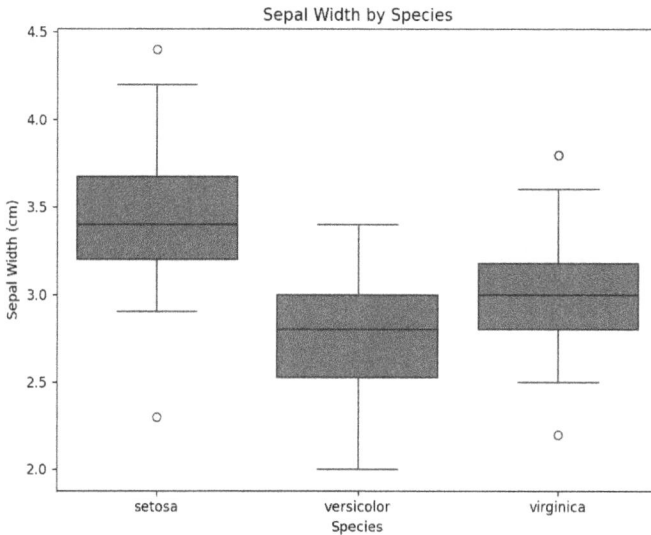

Figure 34. Box Plot of Iris Species

This box plot shows how **sepal width** varies between the three Iris species. You can observe differences in the median values, the spread of the data (interquartile range), and the presence of any outliers.

Step 6: Violin Plots for Feature Distribution and Density

Violin plots combine the features of a box plot with a density plot, showing both the distribution of the data and the probability density.

6.1: Petal Length by Species

```
1  # Violin plot of petal length by species
2  plt.figure(figsize=(8, 6))
3  sns.violinplot(x='species', y='petal_length',
4      data=iris_df)
5  plt.title('Petal Length by Species')
6  plt.xlabel('Species')
7  plt.ylabel('Petal Length (cm)')
8  plt.show()
```

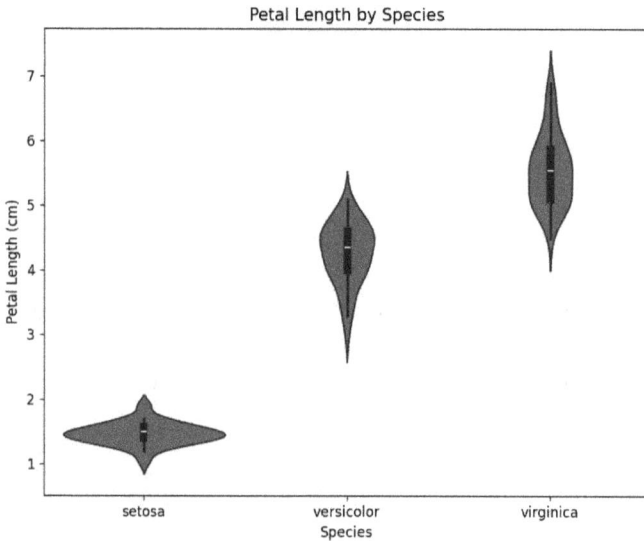

Figure 35. Violin Iris Species Plot

This violin plot shows the distribution and density of **petal length** for each species. You can see not only the range of petal lengths but also how frequent different petal lengths are within each species.

Conclusion

Through this activity, you've learned how to use various visualization techniques—histograms, scatter plots, pair plots, box plots, and violin plots—to explore and visualize the **Iris dataset**. Each type of visualization provides different insights into the relationships and distributions of features in the dataset. These visualizations help highlight patterns that are useful when building machine learning models or gaining a deeper understanding of the data.

In the next section, we will dive into **exploratory data analysis (EDA)** and feature engineering, using these visualizations as a foundation for deeper insights.

Here are some reflective questions and Python challenges focused on the Matplotlib and Seaborn tools from Chapter 5, with an emphasis on visualizing the Iris dataset:

Reflective Questions

1. **Library Functions:**

   - How do Matplotlib and Seaborn complement each other in data visualization tasks?

2. **Visual Customization:**

   - Discuss how the customizability of Matplotlib can be used to refine visual presentations for more complex datasets.

3. **Statistical Insights:**

   - Why might Seaborn be more suitable than Matplotlib for quickly visualizing statistical relationships?

4. **Practical Applications:**

   - Can you think of a scenario where you would prefer `Matplotlib` over `Seaborn`, or vice versa?

5. **Learning Curve:**

   - Reflect on the learning curve associated with `Matplotlib` and `Seaborn`. Which library do you find more intuitive, and why?

6. **Visualization Choices:**

   - How would you decide whether to use a scatter plot, line plot, or histogram for a given set of data?

7. **Theme Utilization:**

   - What advantages do built-in themes in `Seaborn` offer when preparing visualizations for a professional presentation?

8. **Tool Integration:**

   - Describe how `Seaborn` integrates with `Pandas` and why this is beneficial for data scientists.

9. **Visual Analysis:**

   - How does visualizing data help in the preliminary analysis before applying any statistical or machine learning techniques?

10. **Future of Visualization Tools:**

    - With the evolution of visualization tools in Python, what future enhancements or new features do you anticipate or hope for in libraries like `Matplotlib` and `Seaborn`?

Python Challenges on the Iris Dataset

1. **Create a Box Plot Using Seaborn:**

   - Generate a box plot that shows the distribution of sepal length across different species in the Iris dataset.

```
1    import seaborn as sns
2    import matplotlib.pyplot as plt
3
4    # Load the Iris dataset
5    iris = sns.load_dataset('iris')
6
7    # Box plot for sepal length
8    # <add your plot code here>
```

2. **Generate Pair Plots for the Iris Dataset:**

   - Use Seaborn to create pair plots to analyze the pairwise relationships between all the features in the Iris dataset, differentiated by species.

```
1    # Generate pair plots
2    # <add your plot code here>
```

3. **Histogram of Petal Widths:**

   - Create a histogram for petal width using Matplotlib and overlay a Kernel Density Estimate (KDE) using Seaborn.

```
1    # create histogram
2       # <add your plot code here>
```

4. Violin Plot of Sepal Width:

- Use Seaborn to make a violin plot that shows the distributions of sepal width for each species in the Iris dataset.

```
1    # create a violin plot
2    # <add your plot code here>
```

5. Scatter Plot of Sepal Length vs. Petal Length:

- Create a scatter plot to explore the relationship between sepal length and petal length, colored by species.

```
1    # create a scatter plot
2    # <add your plot code here>
```

Chapter 6: Introduction to Exploratory Data Analysis

In any data science project, understanding your data is the critical first step toward meaningful insights. One of the most effective ways to achieve this is through **Exploratory Data Analysis (EDA)**. EDA is a key process used to analyze data sets by summarizing their main characteristics, often using visual methods. This chapter delves into the importance of EDA, how it helps uncover patterns, spot anomalies, test hypotheses, and check assumptions—providing the groundwork for more sophisticated analysis.

What is EDA?

Exploratory Data Analysis (EDA) is the first, crucial step in the data analysis process. It involves employing a variety of statistical techniques and visual tools to explore and summarize the data set, with the goal of understanding its main characteristics without making any assumptions. EDA provides an open-ended approach to data exploration, allowing analysts to discover patterns, relationships, and anomalies that may not be immediately apparent.

EDA can be broken down into two key types:

1. **Quantitative EDA**: This includes summary statistics such as measures of central tendency (mean, median, mode), variability (variance, standard deviation), and the distribution of

the data. It helps to quickly assess the numerical properties of the data set.

2. **Graphical EDA**: Visualizations, such as histograms, box plots, scatter plots, and correlation matrices, play a central role in understanding the distribution of data, relationships between variables, and potential outliers. Graphical methods often reveal patterns or trends that aren't obvious through summary statistics alone.

EDA also serves other vital purposes:

- **Data Cleaning**: It often uncovers missing, inconsistent, or outlier values that need to be addressed before moving forward with analysis.
- **Hypothesis Generation**: EDA is used to generate hypotheses and intuitively understand the data. While formal hypothesis testing is more structured, EDA helps frame initial ideas and guide the formulation of these hypotheses.
- **Assumption Testing**: In statistical modeling, assumptions like normality, linearity, or homoscedasticity are important. EDA helps to check these assumptions early in the process.

Overall, EDA helps ensure that the data is prepared and well-understood before applying any formal models, reducing the risk of errors, and guiding the entire data science pipeline in the right direction.

Steps in EDA

The process of Exploratory Data Analysis (EDA) is a systematic approach to understand the data before moving on to modeling or making predictions. EDA involves several key steps to summarize and interpret the data effectively. Below is a breakdown of the essential steps:

1. Summary Statistics

The first step in EDA is to compute summary statistics, which provide a high-level overview of the data. These statistics describe the central tendencies and variability within the dataset. Common summary statistics include:

- **Mean, Median, Mode**: Measures of central tendency that tell you where most of your data points fall.
- **Range, Variance, and Standard Deviation**: These indicate the spread or variability of the data, helping you understand how dispersed the values are.
- **Minimum and Maximum Values**: These reveal the data's boundaries and can help spot any outliers or extreme values.
- **Skewness and Kurtosis**: These metrics assess the asymmetry and the heaviness of the tails in the data distribution, respectively.

By examining these summary statistics, you gain an immediate understanding of the general characteristics of your dataset, such as whether the data is normally distributed or skewed in some way.

2. Data Visualization

Visual exploration is an essential part of EDA as it helps uncover hidden patterns and trends that may not be apparent in numerical summaries. Common visual tools include:

- **Histograms**: Useful for understanding the distribution of a single variable, showing how frequently each value occurs.
- **Box Plots**: Help identify outliers and visualize the distribution and spread of the data, especially for comparing across different groups.
- **Scatter Plots**: Ideal for examining relationships between two variables and spotting correlations or trends.

- **Correlation Matrices**: Visualize the correlation coefficients between multiple variables to identify linear relationships.

Visual methods offer a clearer, more intuitive understanding of trends, outliers, and relationships between variables that summary statistics alone might not reveal.

3. Finding Trends and Patterns

Once summary statistics and basic visualizations are reviewed, the next step is to dive deeper into identifying trends and patterns within the data. This might include:

- **Identifying Relationships**: Use scatter plots or correlation analysis to see if there are any relationships between variables (e.g., does one variable increase as another increases?).
- **Detecting Seasonality**: In time series data, look for repeating patterns over time, such as daily, weekly, or yearly trends.
- **Spotting Outliers**: Outliers can significantly skew your analysis. EDA helps you identify extreme values or anomalies that may need to be addressed.
- **Examining Group Differences**: Use bar plots, box plots, or group-based statistics to compare different categories or groups within the data (e.g., comparing male vs. female income levels).

4. Handling Missing Data

During EDA, missing data is often encountered. Addressing this is crucial, as missing values can impact analysis and modeling. Techniques include:

- **Identifying Missing Data**: Check how much and where missing data occurs.

- **Imputation**: Filling in missing values with mean, median, or mode, or using more advanced imputation techniques.
- **Omission**: Removing rows or columns with a high amount of missing data when imputation isn't appropriate.

5. Testing Assumptions

Before moving forward with statistical models, you should test whether the data meets the assumptions required for these models. EDA provides an opportunity to:

- **Test Normality**: Using Q-Q plots or histograms to check if data is normally distributed.
- **Check for Homoscedasticity**: Ensuring that the variance is consistent across variables.
- **Linearity**: Verifying if relationships between variables are linear, which is important for many models.

By following these steps in EDA, you can better understand your data, uncover valuable insights, and ensure that the data is clean and ready for further analysis or modeling. This foundational process is key to minimizing errors and ensuring the success of subsequent steps in your data analysis project.

Case Study: Performing EDA on a Dataset

To illustrate the process of Exploratory Data Analysis (EDA), let's walk through a case study using the well-known *House Prices* dataset from Kaggle. This dataset contains information about various factors that influence housing prices, including features like the number of bedrooms, square footage, and neighborhood. We will use EDA techniques to explore the dataset and gain insights into the data, ultimately guiding further analysis.

Step 1: Load and Inspect the Data

The first step in EDA is loading the data and taking a preliminary look at its structure.

```
import pandas as pd

# Load the dataset
data = pd.read_csv('house_prices.csv')

# Inspect the first few rows
data.head()
```

Out[5]:

	date	price	bedrooms	bathrooms	sqft_living	sqft_lot	floors	waterfront	view	condition	sqft_above	sqft_basement	yr_built	yr_renovated	stre
0	2014-05-09 00:00:00	376000.0	3.0	2.00	1340	1384	3.0	0	0	3	1340	0	2008	0	924 92 Fremc Ave
1	2014-05-09 00:00:00	800000.0	4.0	3.25	3540	159430	2.0	0	0	3	3540	0	2007	0	330 NE 24
2	2014-05-09 00:00:00	2238888.0	5.0	6.50	7270	130017	2.0	0	0	3	6420	850	2010	0	70 270th
3	2014-05-09 00:00:00	324000.0	3.0	2.25	998	904	2.0	0	0	3	798	200	2007	0	820 N 95th
4	2014-05-10 00:00:00	549900.0	5.0	2.75	3060	7015	1.0	0	0	5	1600	1460	1979	0	108 3 Ave S

Figure 36. First 5 Rows of House Prices Dataset

This provides a quick view of the columns and data types. In this dataset, we have columns like `SalePrice` (the target variable), `LotArea`, `OverallQual` (overall quality), `YearBuilt`, and many more that describe the features of each house.

Step 2: Summary Statistics

Next, we compute summary statistics to understand the distribution of numerical variables, including key measures like mean, median, and variance.

```
1  # Summary statistics for numerical columns
2  data.describe()
```

Out[6]:

	price	bedrooms	bathrooms	sqft_living	sqft_lot	floors
count	4.140000e+03	4140.000000	4140.000000	4140.000000	4.140000e+03	4140.000000
mean	5.530629e+05	3.400483	2.163043	2143.638889	1.469764e+04	1.514130
std	5.836865e+05	0.903939	0.784733	957.481621	3.587684e+04	0.534941
min	0.000000e+00	0.000000	0.000000	370.000000	6.380000e+02	1.000000
25%	3.200000e+05	3.000000	1.750000	1470.000000	5.000000e+03	1.000000
50%	4.600000e+05	3.000000	2.250000	1980.000000	7.676000e+03	1.500000
75%	6.591250e+05	4.000000	2.500000	2620.000000	1.100000e+04	2.000000
max	2.659000e+07	8.000000	6.750000	10040.000000	1.074218e+06	3.500000

Figure 37. Summary Statistics of Housing Prices

From the summary, we can learn key insights such as:

- The **mean sale price** of homes is $553,063.
- The **minimum sale price** is $0, while the **maximum** is $26,000,000, indicating a wide range of home prices.
- Features like squarefoot living and squarefoot lot also have high variability, hinting at the potential importance of these features.

Step 3: Data Visualization

Visualizations help reveal trends and patterns that might not be obvious from summary statistics. Below are some useful visualizations for this dataset:

- **Histogram for `price`:** To examine the distribution of house prices.

```
import matplotlib.pyplot as plt

# Plot histogram of SalePrice
plt.hist(data['price'], bins=50, color='blue')
plt.title('Distribution of Sale Prices')
plt.xlabel('Sale Price')
plt.ylabel('Frequency')
plt.show()
```

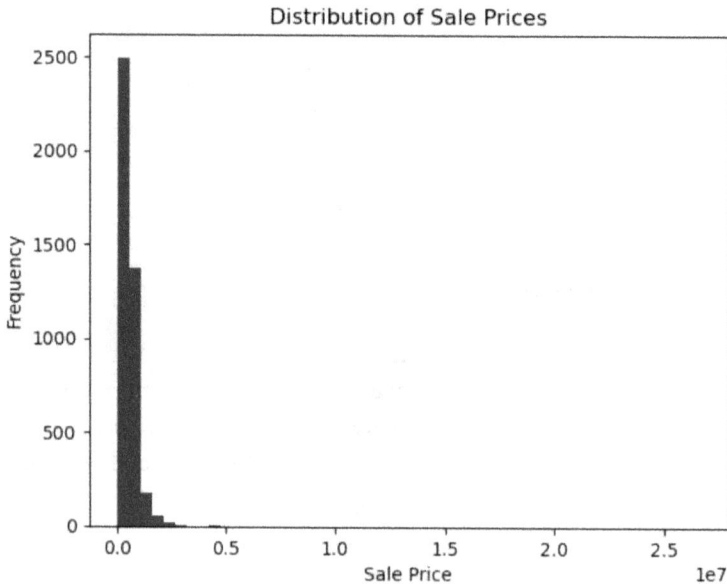

Figure 38. Distribution of Sales Price

This histogram may show that house prices are skewed to the right, indicating more lower-priced homes and fewer high-end homes.

- **Scatter plot between `square foot living` and `SalePrice`**: To check the relationship between ground living area and house prices.

```
1  # Scatter plot for GrLivArea vs SalePrice
2  plt.scatter(data['sqft_living'], data['price'])
3  plt.title('Sale Price vs Ground Living Area')
4  plt.xlabel('Ground Living Area (sq ft)')
5  plt.ylabel('Sale Price')
6  plt.show()
```

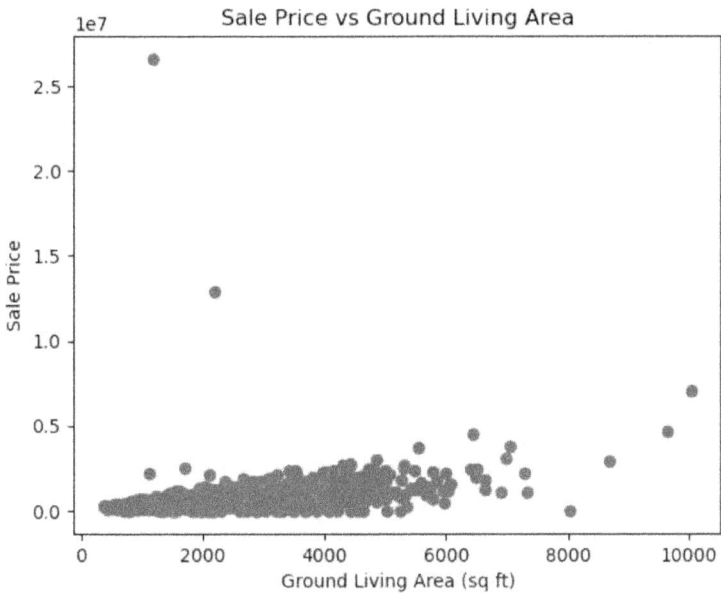

Figure 39. Scatter Plot of Living Area vs Sales Price

In this plot, you might notice a positive correlation between living area and sale price: as the size of the home increases, so does its price. This is a potential predictor for building future models.

Step 4: Detecting Trends and Patterns

Next, we look for trends and patterns that could be useful in predicting house prices:

- **Correlation Matrix**: To identify relationships between numerical variables, we compute a correlation matrix.

```python
# Correlation matrix
corr_matrix = data.corr()
import seaborn as sns
# Increase the figure size for better readability
plt.figure(figsize=(12, 8))

# Plot the heatmap with rounded annotations and
# adjust the color map
sns.heatmap(corr_matrix, annot=True, fmt='.2f',
    cmap='coolwarm', linewidths=0.5)

plt.title('Correlation Matrix')
plt.show()
```

Correlation Matrix

	price	bedrooms	bathrooms	sqft_living	sqft_lot	floors	waterfront	view	condition	sqft_above	sqft_basement	yr_built	yr_renovated
price	1.00	0.19	0.32	0.42	0.05	0.14	0.13	0.22	0.03	0.36	0.20	0.03	-0.03
bedrooms	0.19	1.00	0.54	0.59	0.07	0.18	-0.00	0.11	0.02	0.48	0.33	0.15	-0.07
bathrooms	0.32	0.54	1.00	0.76	0.10	0.49	0.08	0.21	-0.12	0.69	0.29	0.47	-0.22
sqft_living	0.42	0.59	0.76	1.00	0.19	0.35	0.13	0.31	-0.07	0.87	0.44	0.30	-0.12
sqft_lot	0.05	0.07	0.10	0.19	1.00	-0.00	0.02	0.07	0.01	0.20	0.02	0.05	-0.02
floors	0.14	0.18	0.49	0.35	-0.00	1.00	0.02	0.03	-0.26	0.52	-0.26	0.47	-0.23
waterfront	0.13	-0.00	0.08	0.13	0.02	0.02	1.00	0.36	0.00	0.08	0.11	-0.03	0.01
view	0.22	0.11	0.21	0.31	0.07	0.03	0.36	1.00	0.06	0.17	0.33	-0.07	0.03
condition	0.03	0.02	-0.12	-0.07	0.01	-0.26	0.00	0.06	1.00	-0.18	0.20	-0.40	-0.19
sqft_above	0.36	0.48	0.69	0.87	0.20	0.52	0.08	0.17	-0.18	1.00	-0.05	0.42	-0.16
sqft_basement	0.20	0.33	0.29	0.44	0.02	-0.26	0.11	0.33	0.20	-0.05	1.00	-0.16	0.05
yr_built	0.03	0.15	0.47	0.30	0.05	0.47	-0.03	-0.07	-0.40	0.42	-0.16	1.00	-0.32
yr_renovated	-0.03	-0.07	-0.22	-0.12	-0.02	-0.23	0.01	0.03	-0.19	-0.16	0.05	-0.32	1.00

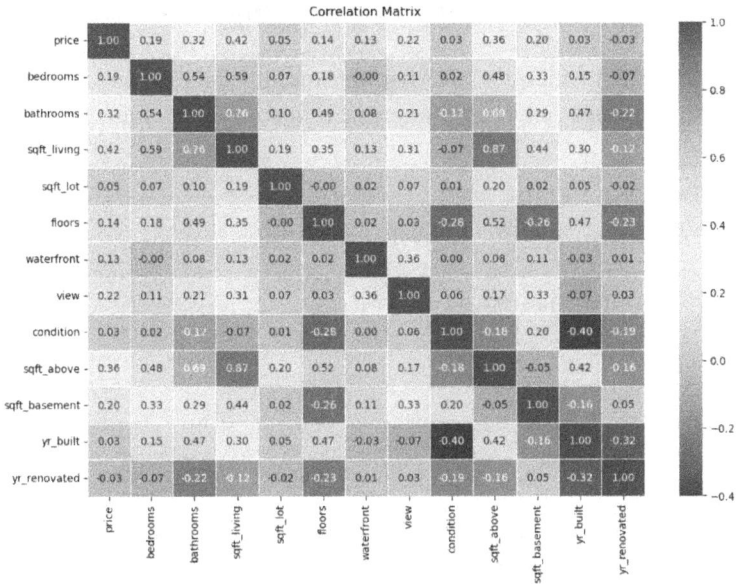

Figure 40. Heat Map of Housing Prices

From the heatmap, you can quickly spot which features are strongly correlated with price. For example, bathrooms (overall quality of the house) and sqft_living might show high positive correlations, while yr_built may also exhibit a notable relationship.

Negative correlations, on the other hand, indicate an inverse relationship between variables—when one increases, the other tends to decrease. For instance, yr_built (the year the house was built) may show a negative correlation with condition. This suggests that older homes tend to be in worse condition compared to newer homes. Similarly, other variables like sqft_basement negatively correlate with floors, meaning homes with lots of floors may make up for not having a basement.

Step 5: Handling Missing Data

It's crucial to identify and handle missing data during EDA, as
missing values can disrupt analysis or lead to inaccurate results.

```
# Check for missing data
missing_data = data.isnull().sum().sort_values
    (ascending=False)
missing_data[missing_data > 0]
```

```
Out[16]:  date               0
          price              0
          statezip           0
          city               0
          street             0
          yr_renovated       0
          yr_built           0
          sqft_basement      0
          sqft_above         0
          condition          0
          view               0
          waterfront         0
          floors             0
          sqft_lot           0
          sqft_living        0
          bathrooms          0
          bedrooms           0
          country            0
          dtype: int64
```

Figure 41. Missing Data By Column

Here's a refined version of your paragraph:

_____

This process generates a list of columns with missing values. In this dataset, we found no missing data. However, if missing data is present in certain columns, there are several strategies you can use to address it:

- **Impute missing values**: This involves replacing missing values with a statistic like the median or mode, which is especially useful when missing data is minimal.
- **Drop columns or rows**: If the proportion of missing data is large, and the feature isn't critical, you can consider dropping the affected columns or rows.

For instance, if we had a column such as PoolQC with missing data, and only a small percentage of homes have pools, we might decide to drop that column entirely if it's not essential to our analysis.

Step 6: Identifying Outliers

Outliers can significantly affect the performance of models, so it's important to identify them during EDA. For instance, scatter plots can be used to check for unusual values:

```
1  # Identify outliers in sqft_living and price
2  plt.scatter(data['sqft_living'], data['price'])
3  plt.title('Identifying Outliers: Sale Price vs
4      Ground Living Area')
5  plt.xlabel('Ground Living Area')
6  plt.ylabel('Sale Price')
7  plt.show()
```

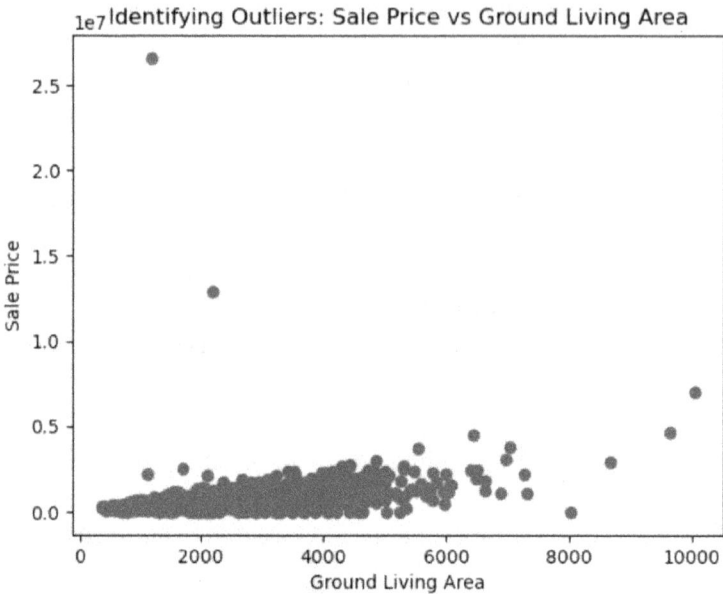

Figure 42. Scatter Plot Shows Outliers

From this plot, you might notice a few outliers where small homes have unusually high sale prices. These outliers could be due to data entry errors, or they might represent special cases like expensive zipcodes. Based on this, you could choose to investigate or remove these records from the analysis.

Removing the Outliers

From the scattere3d, we can see the outliers that are interfering with our analysis and they fall above a sales price above 10 million, let's take a look at those records:

```
1  data[data['price'] > 10000000]
```

```
Out[20]:
```

	date	price	bedrooms	bathrooms	sqft_living	sqft_lot	floors
3886	2014-06-23 00:00:00	12899000.0	3.0	2.5	2190	11394	1.0
3890	2014-07-03 00:00:00	26590000.0	3.0	2.0	1180	7793	1.0

Figure 43. Price Outliers

These seem extremely high for a square foot living space of 1000-2000 square feet, 1 floor, and on much less than an acre. Seems like the entries could be typos. We'll remove these two records and then recheck our distribution.

```
1  # remove the outliers
2  data = data[data['price'] < 10000000]
3
4  # check they are removed
5  data[data['price'] > 10000000]
```

Now if we look at our distribution plot, its no longer squeezed into a small width area on the graph:

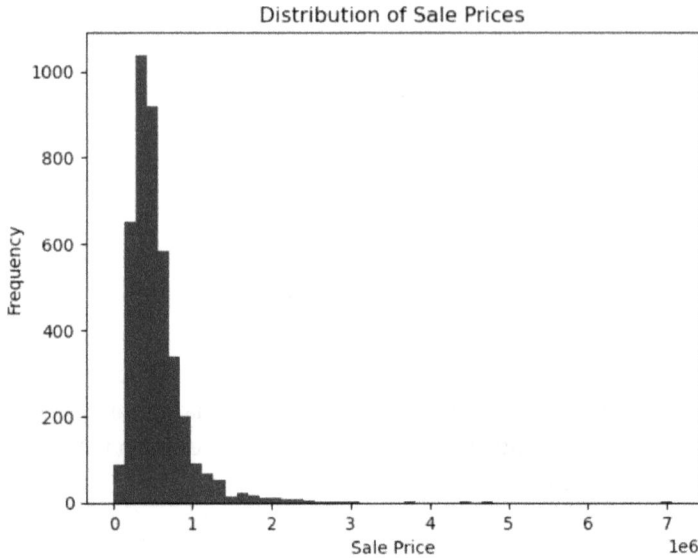

Figure 44. Distribution Plot After Removing Outliers

We can also see more clearly the price distribution and the skewing of sales prices to the right.

Step 7: Testing Assumptions

Lastly, we check whether the data meets the assumptions of normality and linearity, which are critical for many statistical models.

- **Normality Test for SalePrice**: Use a Q-Q plot to check if house prices follow a normal distribution.

```
1   import scipy.stats as stats
2
3   # Q-Q plot for SalePrice
4   stats.probplot(data['price'], dist="norm",
5       plot=plt)
6   plt.title('Q-Q Plot of Sale Prices')
7   plt.show()
```

Figure 45. Q-Q Plot of Price

If the points deviate significantly from the straight line, it suggests that house prices are not normally distributed, which may require transformations (like a log transformation) in future analyses.

Conclusion

In this case study, we walked through the essential steps of Exploratory Data Analysis using the *House Prices* dataset. By using summary statistics, visualizations, and detecting patterns, we gained insights into the key variables that influence house prices. Additionally, handling missing data and identifying outliers ensured the dataset was clean and ready for further analysis or predictive modeling. EDA is a powerful process that uncovers the underlying structure of the data, allowing for informed decisions and better model performance.

Below are some reflective questions for Chapter 6 on Exploratory Data Analysis (EDA):

1. **Understanding EDA's Role:**

   - How does EDA facilitate a better understanding and treatment of data before more formal and complex analyses are performed?

2. **Quantitative vs. Graphical EDA:**

   - Discuss the benefits and limitations of quantitative EDA compared to graphical EDA. Can you think of scenarios where one might be more useful than the other?

3. **Importance of Data Cleaning:**

   - Reflect on the impact of missing, inconsistent, or outlier values discovered during EDA. How can these affect the outcomes of data analysis?

4. **Hypothesis Generation:**

   - How does EDA aid in hypothesis generation? Provide an example based on your understanding.

5. **Assumption Testing:**

   - Why is testing assumptions an integral part of EDA, especially before proceeding to complex data modeling?

6. **Tools and Techniques:**

   - What are some of the most crucial tools or techniques in EDA that every data scientist should master? Why?

7. **Challenges in EDA:**

   - What challenges might you face while performing EDA and how can you overcome them?

8. **EDA in Different Domains:**

   - Can the approach to EDA vary by industry or type of data (e.g., time-series vs. cross-sectional data)? Provide examples.

9. **Integrating EDA with Other Data Processes:**

   - How does EDA integrate with other processes in data science projects, such as data cleaning and feature engineering?

10. **Future of EDA:**

    - With advancements in AI and machine learning, how do you envision the future of EDA? Will automated EDA tools replace traditional methods?

Python Challenges for EDA on a Dataset

Let's propose challenges that utilize the well-known Iris dataset to explore different aspects of EDA.

1. **Basic Summary Statistics and Visualization:**

   - Generate summary statistics (mean, median, mode) for each feature in the Iris dataset.
   - Create histograms for each feature to visualize their distributions.

```
1    import seaborn as sns
2    import matplotlib.pyplot as plt
3
4    # Load the dataset
5    iris = sns.load_dataset('iris')
6
7    # Display summary statistics
8    # <add your print statement here>
9
10   # Plot histograms
11   # <draw the histogram here>
12   plt.show()
```

2. **Explore Relationships Between Features:**

   - Use scatter plots to examine the relationships between each pair of features.
   - Create a correlation matrix to identify the relationships quantitatively.

```
1    # Scatter plot matrix
2    # hint: use a pair plot
3    plt.show()
4
5    # Correlation matrix heatmap
6    plt.figure(figsize=(8, 6))
7    # <draw your heatmap here>
8    plt.show()
```

3. **Identify Outliers Using Box Plots**:

   - Generate box plots for each feature to visualize potential outliers.

```
1    # Box plots to identify outliers
2    plt.figure(figsize=(10, 8))
3    # <add boxplots here>
4    plt.title('Box plot for detecting outliers in Iris datase\
5    t')
6    plt.show()
```

4. **Handle Missing Data**:

   - Artificially introduce missing values in the Iris dataset, then apply imputation methods.
   - **Hint**: Use iris.loc (where iris is a dataframe) and assign a subset of columns and rows to np.nan

```
1    import numpy as np
2    # Introduce missing values
3    # <add your code here>
4    print(iris.isnull().sum())
5    # Impute missing values with the mean
6    # <add your code here>
7    print(iris.isnull().sum())
```

5. **Testing Assumptions**:

   - Perform a normality test using Q-Q plots for the 'sepal_-width' feature.

```
1    import scipy.stats as stats
2
3    # Q-Q plot for normality test
4    # <add your code here>
5    plt.title('Q-Q Plot for Sepal Width')
6    plt.show()
```

Chapter 7: What is Machine Learning?

Machine Learning (ML) has become one of the most transformative technologies in the modern world, changing how industries, governments, and individuals process and utilize data. At its core, ML is a subset of artificial intelligence (AI) that enables systems to learn from data, identify patterns, and make decisions with minimal human intervention. This chapter introduces the foundational concepts of machine learning, its types, and how it serves as the backbone for numerous applications across various fields.

Definition of Machine Learning

Machine learning can be defined as a method of data analysis that automates analytical model building. It's a science that allows computers to perform tasks without being explicitly programmed. In the words of Arthur Samuel, one of the early pioneers in machine learning, it is "the field of study that gives computers the ability to learn without being explicitly programmed."

In practical terms, machine learning involves developing algorithms that can process vast amounts of data, discern patterns, and make predictions based on new data inputs. This ability to learn and adapt from experience differentiates ML from traditional programming approaches.

Significance of Machine Learning

Machine learning has achieved widespread significance due to its unique ability to derive insights from data and apply them to improve processes, drive automation, and make informed predictions. Its importance is evident in the following ways:

1. **Enhanced Decision-Making**: ML helps organizations make data-driven decisions that are faster, more accurate, and often predictive, allowing for real-time insights.
2. **Scalability**: Traditional analytics methods struggle to manage and process the vast amounts of data generated daily. ML algorithms are built to scale with big data, analyzing complex patterns across millions of records.
3. **Adaptability**: Machine learning models evolve and improve as they are exposed to more data over time, enhancing their predictive accuracy and adaptability to changes.
4. **Automation**: With ML, tasks traditionally requiring human intervention can now be automated, freeing up resources for more strategic roles and reducing operational costs.
5. **Personalization**: ML enables hyper-personalization in products and services, making it a cornerstone of customer-centric businesses. Netflix's recommendation engine, Amazon's personalized shopping experience, and personalized marketing emails are examples of ML-powered personalization in action.

Machine Learning in Everyday Life

The reach of machine learning extends into various domains, making it a technology that individuals frequently encounter, often unknowingly. Some applications include:

- **Social Media Algorithms**: Platforms like Facebook, Instagram, and TikTok use ML algorithms to personalize user feeds, suggest friends, and display targeted ads.
- **E-commerce Recommendations**: Retailers like Amazon and eBay use machine learning to analyze user preferences and recommend products.
- **Healthcare**: Machine learning assists doctors in diagnosing diseases by analyzing patient data, identifying patterns in medical imaging, and even predicting potential health risks.
- **Finance**: ML is used for fraud detection, risk assessment, and automated trading, enhancing security and operational efficiency.
- **Self-Driving Cars**: Autonomous vehicles rely on ML models to process data from sensors and make real-time decisions about speed, direction, and collision avoidance.

Types of Machine Learning

Machine learning can be broadly categorized into three types: supervised learning, unsupervised learning, and reinforcement learning.

1. **Supervised Learning**: In supervised learning, models are trained on a labeled dataset, meaning that each training example is paired with an output label. The model learns to make predictions based on this labeled data. This type of learning is commonly used for tasks like image classification, spam detection, and predicting stock prices.
2. **Unsupervised Learning**: In contrast to supervised learning, unsupervised learning algorithms work on unlabeled data, where the model tries to find patterns or groupings on its own. Clustering, where data points are grouped into clusters,

is a common application. Unsupervised learning is useful for customer segmentation, anomaly detection, and pattern recognition.

3. **Reinforcement Learning**: This is a type of machine learning where an agent interacts with an environment and learns by receiving rewards or penalties for actions taken. Reinforcement learning is widely used in robotics, gaming, and navigation.

Hands-On Example: Predicting Housing Prices with Python

To make machine learning concrete, let's walk through a simple example using Python to predict housing prices. In this example, we'll use a small dataset with features such as the number of rooms and the age of a house to predict its price. We'll leverage the `scikit-learn` library, which provides ready-to-use machine learning tools.

Step 1: Install Scikit-Learn

If you don't already have `scikit-learn`, install it by running:

```
!pip install scikit-learn
```

Step 2: Import Libraries and Set Up Data

Here, we'll use the following imports:

- `train_test_split` to split our data for training and testing.
- `LinearRegression`, a basic machine learning model for predictions.

```
1  from sklearn.model_selection import
2      train_test_split
3  from sklearn.linear_model import LinearRegression
4  import numpy as np
5
6  # Sample data: [number of rooms, age of house]
7  X = np.array([[3, 20], [2, 15], [4, 5], [3, 30],
8      [5, 10]])
9  # Prices in thousands of dollars
10 y = np.array([150, 120, 200, 130, 250])
```

Step 3: Split the Data

We divide the data into a training set (to teach the model) and a test set (to evaluate its predictions).

```
1  X_train, X_test, y_train, y_test =
2      train_test_split(X, y, test_size=0.2,
3      random_state=42)
```

Step 4: Train the Model

Next, we initialize a LinearRegression model and train it with our training data.

```
1  model = LinearRegression()
2  model.fit(X_train, y_train)
```

Step 5: Make Predictions and Evaluate the Model

Now that our model is trained, we can use it to predict housing prices for our test set. Then, we'll compare these predictions with the actual values.

```
1  predictions = model.predict(X_test)
2  print("Predicted prices:", predictions)
3  print("Actual prices:", y_test)
```

Output:

```
1  Predicted prices: [100]
2  Actual prices: [120]
```

The output shows the predicted and actual prices, which may be close if the model is trained well.

Step 6: Visualizing the Results

Finally we can visualize the how well the model fits the data by drawing the actual points and then plotting the linear model.

```
1  # Predicting values for the line plot
2  # Assuming the second feature is a constant
3  # (e.g., 1 for intercept simulation)
4  import numpy as np
5  import matplotlib.pyplot as plt
6
7  # Generate prediction range for one feature
8  X_range = np.linspace(1, 6, 100).reshape(-1, 1)
9
10 # Add second feature if needed (e.g., constant
11 # feature)
12 second_feature = np.ones((X_range.shape[0], 1))
13 X_range = np.hstack((X_range, second_feature))
14
15 # Predict values
16 y_range_pred = model.predict(X_range)
17
18 # Plotting
19 plt.figure(figsize=(8, 6))
```

```
20   plt.scatter(X_train[:, 0], y_train, color='blue',
21       label="Training Data")
22   # Assuming first feature is used
23   plt.scatter(X_test[:, 0], y_test, color='red',
24       label="Test Data")
25   plt.plot(X_range[:, 0], y_range_pred,
26       color='green', linestyle="--",
27       label="Regression Line")
28   plt.xlabel("Number of Rooms")
29   plt.ylabel("House Price (in $1000s)")
30   plt.title("Linear Regression on Housing Data
31       (Rooms vs. Price)")
32   plt.legend()
33   plt.grid(True)
34   plt.show()
```

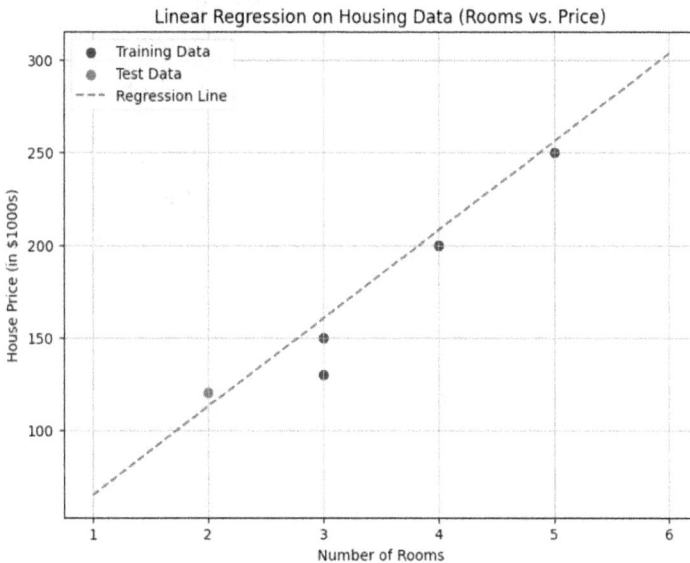

Figure 46. Plot of Actual Vs Predicted Model

Step 7: Evaluating Model Error After making predictions, it's essential to assess how accurately our model performed by measuring the error. One common metric for this is the Mean Absolute Error (MAE), which gives us the average absolute difference between the predicted and actual prices.

```
1   from sklearn.metrics import mean_absolute_error
2
3   # Calculate Mean Absolute Error
4   mae = mean_absolute_error(y_test, predictions)
5   print("Mean Absolute Error:", mae)
```

Output Mean Absolute Error: 20.000000000000014

The Mean Absolute Error provides a sense of how much, on average, our model's predictions deviate from the actual prices. A lower MAE indicates a more accurate model, meaning it's better at making predictions on unseen data.

In our example, this MAE value gives insight into how much we can expect our predicted housing prices to differ from the actual prices in this dataset. Evaluating error is crucial as it allows us to understand the model's reliability and where there may be room for improvement.

Explanation of the Code

1. **Data Preparation**: We create two arrays: X contains our features (rooms and age), and y contains target prices.
2. **Splitting the Data**: We split X and y into training and testing sets, ensuring the model has separate data to learn and be evaluated on.

3. **Model Training**: The `LinearRegression` model is initialized and trained with our training data. It learns relationships between features and target prices.

4. **Making Predictions**: We use the trained model to predict the prices in the test set and then print out the results.

5. **Visualizing Our Results**: To better understand how well our model fits the data, we can visualize the relationship between the number of rooms and predicted prices by plotting the training and test data points, alongside the model's prediction line. This line represents our linear regression model's attempt to capture the underlying trend in the data, helping us see where our predictions align with actual prices.

6. **Evaluating Model Error**: After making predictions, we assess the model's accuracy by calculating the Mean Absolute Error (MAE), which gives us the average absolute difference between predicted and actual prices. A lower MAE suggests the model is better at making accurate predictions. By evaluating error, we gain insights into the model's performance and identify areas for potential improvement.

Conclusion

In this chapter, we introduced the concept of machine learning, covering its definition, significance, and primary types—supervised, unsupervised, and reinforcement learning. Machine learning is not just a technical field but a transformative technology that powers everything from predictive analytics to personal recommendations and autonomous systems. By exploring real-world applications, we saw how machine learning models are already part of our daily lives.

The hands-on activity demonstrated how we can use Python to build a simple predictive model, using housing data to forecast prices. We walked through each step of the process: preparing and splitting data, training a model, making predictions, visualizing results, and evaluating model error. This example underscores the simplicity and practicality of implementing basic machine learning concepts and highlights the importance of testing and improving models to achieve reliable results.

As we move forward, we will delve deeper into specific machine learning algorithms, covering how they work, their use cases, and how to evaluate and optimize them. With a solid foundation in the basics, you're now prepared to explore machine learning's vast potential and its powerful applications in data analytics.

Here are some reflective questions for Chapter 7 on Machine Learning:

Reflective Questions

1. **Understanding Machine Learning**:

   - How would you explain the concept of machine learning to someone without a technical background?

2. **Impact of Machine Learning**:

   - What are some significant impacts of machine learning you have noticed in everyday life? Can you identify any potential negative impacts?

3. **Machine Learning vs. Traditional Programming**:

   - How does machine learning differ from traditional programming methods? What advantages does machine learning offer over traditional approaches?

4. **Machine Learning Categories**:

   - Explain the differences between supervised, unsupervised, and reinforcement learning. Can you provide real-world examples for each category?

5. **Challenges in Machine Learning**:

   - What are some of the biggest challenges faced when working with machine learning models?

6. **Ethical Considerations**:

   - Discuss the ethical considerations that should be taken into account when developing machine learning systems.

7. **Future of Machine Learning**:

   - How do you envision the future of machine learning evolving over the next decade? What roles do you think machine learning will play in future technologies?

8. **Machine Learning in Industries**:

   - How is machine learning transforming different industries? Provide examples from at least two different sectors.

9. **Skill Development**:

   - What skills do you think are essential for someone interested in pursuing a career in machine learning?

10. **Machine Learning Tools and Technologies**:

    - What are some of the key tools and technologies that are indispensable for practitioners in the field of machine learning?

Python Challenges for Understanding Machine Learning:

1. **Building a Simple Classifier:**

   - Use scikit-learn to create a simple classifier to predict whether a person buys a product based on features such as age and salary.

```
1    from sklearn.datasets import make_classification
2    from sklearn.model_selection import train_test_split
3    from sklearn.ensemble import RandomForestClassifier
4    from sklearn.metrics import accuracy_score
5
6    # Generate a synthetic dataset
7    X, y = make_classification(
8        n_samples=1000,
9        n_features=2,
10       n_informative=2,
11       n_redundant=0,
12       random_state=42)
13
14   # Split the data
15   X_train, X_test, y_train, y_test =
16     train_test_split(X, y, test_size=0.2,
17     random_state=42)
18
19   # Train a Random Forest classifier
20   # <Add your training code here>
21
22
23   # Predict and evaluate the model on the test data
24   predictions = model.predict(X_test)
25   accuracy = accuracy_score(y_test, predictions)
26   print(f'Accuracy: {accuracy:.2f}')
```

2. **Clustering with K-Means:**

- Apply K-Means clustering to group similar data points together. Use the Iris dataset to cluster the data based on features.

```
1   from sklearn.cluster import KMeans
2   import seaborn as sns
3
4   # Load the Iris dataset
5   iris = sns.load_dataset('iris')
6   features = iris[['sepal_length',
7       'sepal_width', 'petal_length', 'petal_width']]
8
9   # Apply K-Means Clustering
10  # <create a kmeans cluster n_custers = 3 here>
11  # <calculate predictions here>
12
13  # Plot the clusters
14  plt.scatter(features['petal_length'], features
15      ['petal_width'], c=predictions)
16  plt.title('K-Means Clustering on Iris Dataset')
17  plt.xlabel('Petal Length')
18  plt.ylabel('Petal Width')
19  plt.show()
```

3. **Regression Analysis:**

- Perform a regression analysis to predict housing prices based on multiple features such as number of rooms, age of the house, and location.

```
1   from sklearn.datasets import fetch_california_housing
2   from sklearn.linear_model import LinearRegression
3   from sklearn.metrics import mean_squared_error
4
5   # Load Boston housing dataset
6   housing = fetch_california_housing()
7   X = housing.data
8   y = housing.target
9
10  # Split data
11  X_train, X_test, y_train, y_test = train_test_split(
12      X, y, test_size=0.25, random_state=42)
13
14  # Train a Linear Regression model
15  # <train the linear regression model here>
16  # <make predictions on the test data>
17
18  # Predict and calculate MSE
19  # <calulate the mean square error here>
20  print(f'Mean Squared Error: {mse:.2f}')
```

Chapter 8: Introduction to Regression and Classification

Supervised learning is one of the most widely used types of machine learning and serves as the foundation for numerous applications, from forecasting and risk assessment to image recognition and sentiment analysis. At its core, supervised learning involves training a model to make predictions based on labeled data. Labeled data means that each input data point has a corresponding output or label, enabling the model to "learn" patterns and relationships. The ultimate goal is to create a model capable of accurately predicting outputs for new, unseen data based on these learned relationships.

Supervised learning tasks fall into two main categories: **regression** and **classification**. Understanding these two categories is essential for anyone interested in building and deploying machine learning models, as they are the basis for solving many real-world problems. In this section, we'll dive into the specifics of regression and classification, discussing the types of data suitable for each, the differences between the two, and the typical use cases where they excel.

Regression: Predicting Continuous Values

Regression is a type of supervised learning used to predict **continuous numerical values**. In other words, regression models estimate a quantity based on one or more features, where the target variable can take any real value within a specified range. Classic examples of regression tasks include predicting house prices, forecasting stock prices, estimating demand for a product, and calculating potential patient outcomes in healthcare.

How Regression Works

At its essence, regression aims to find a mathematical relationship between input features and the target variable. The relationship is represented as a function, often in the form of a line or curve, that can approximate or model the data points in a way that minimizes the error between predicted values and actual values. In its simplest form, regression is linear, meaning it tries to fit a straight line to the data. However, it can also be polynomial or take on other complex forms if the relationship between features and the target variable is non-linear.

The general approach to building a regression model involves:

1. **Collecting and Preparing Data**: Collecting a dataset with both input features and corresponding target values, then cleaning and preprocessing this data to ensure it's in a suitable format.
2. **Selecting a Regression Model**: Choosing a regression algorithm that fits the problem's nature and complexity, such as linear regression for simple linear relationships or polynomial regression for more complex patterns.

3. **Training the Model**: Using a subset of the data (called the training set) to allow the model to "learn" by adjusting its parameters to minimize prediction errors.

4. **Evaluating the Model**: Testing the model on a different subset of the data (the test set) to assess its performance and measure how well it generalizes to new data.

Evaluation Metrics in Regression

To gauge how well a regression model performs, we rely on evaluation metrics that quantify the error between predicted and actual values. Some of the common metrics include:

- **Mean Absolute Error (MAE)**: The average absolute difference between predicted and actual values. It is intuitive and easy to interpret, representing the average error in the units of the target variable.

- **Mean Squared Error (MSE)**: The average of the squared differences between predicted and actual values. Squaring the errors penalizes larger discrepancies, making MSE sensitive to outliers.

- **Root Mean Squared Error (RMSE)**: The square root of MSE, which makes it comparable to the target variable's units. It's often preferred when dealing with outliers in the data.

- **R-squared**: A statistical measure that indicates how much of the variance in the target variable is explained by the model. R-squared values range from 0 to 1, with higher values indicating a better fit.

Using these metrics, we can analyze the accuracy of a regression model and determine whether it's suitable for deployment.

When to Use Regression

Regression is particularly useful when the target variable is continuous and the relationship between input features and the target variable can be approximated mathematically. It's ideal for tasks where trends need to be forecasted, such as predicting temperatures, assessing credit risk, or estimating economic growth.

Classification: Predicting Categorical Outcomes

In contrast to regression, **classification** is used to predict **discrete, categorical outcomes**. The target variable in classification tasks can only take on specific labels or classes, such as "spam" or "not spam," "defective" or "non-defective," or multiple classes, such as predicting which type of plant species a given flower belongs to. Classification models output probabilities for each class, which helps in determining the likelihood of a data point belonging to a specific category.

How Classification Works

Classification models learn patterns and relationships by analyzing labeled examples of each class. For instance, in an email spam classifier, the model learns from emails marked as "spam" or "not spam." It identifies features such as the frequency of certain words, email structure, or sender domain, which are often associated with each class. When new emails are introduced, the model assigns a probability for each class based on learned patterns, ultimately predicting whether the email is likely spam.

The steps in building a classification model are similar to those in regression:

1. **Data Preparation:** Collect and preprocess a labeled dataset, ensuring the features and target labels are suitable for analysis.
2. **Choosing a Classification Algorithm:** Depending on the problem complexity, select an appropriate algorithm, such as logistic regression for binary outcomes or decision trees for more intricate relationships.
3. **Training the Model:** Use the training data to allow the model to learn the associations between input features and classes, adjusting model parameters to minimize classification error.
4. **Evaluating Model Performance:** Evaluate the model on test data to understand its classification accuracy and error rate.

Evaluation Metrics in Classification

In classification tasks, different metrics help assess how well the model distinguishes between classes:

- **Accuracy:** The percentage of correctly classified instances out of the total number of instances. It's straightforward but can be misleading in imbalanced datasets.
- **Precision and Recall:** Precision is the percentage of correctly classified positive instances, while recall is the percentage of actual positive instances that were correctly identified. They are particularly useful in contexts like medical diagnosis or fraud detection, where one class is more important.
- **F1 Score:** The harmonic mean of precision and recall, balancing both metrics to provide a single measure of performance, especially when data is imbalanced.
- **Confusion Matrix:** A table showing the number of correct and incorrect predictions for each class. It offers a detailed breakdown of how the model performs across each class, highlighting misclassifications.

When to Use Classification

Classification is the go-to approach when the target variable is categorical. It is widely used in fields where decisions are binary or involve multiple distinct categories. For example, classification is critical in:

- **Healthcare**: Diagnosing diseases based on patient symptoms or imaging.
- **Finance**: Flagging fraudulent transactions.
- **Customer Service**: Analyzing customer sentiment and categorizing feedback.

Comparing Regression and Classification

While both regression and classification involve learning from labeled data, they serve different purposes and require distinct types of evaluation. Here's a summary of their key differences:

Aspect	Regression	Classification
Target Variable	Continuous (e.g., house price)	Categorical (e.g., spam/not spam)
Goal	Predict a quantity	Predict a category
Common Algorithms	Linear regression, polynomial regression	Logistic regression, decision trees
Evaluation Metrics	MAE, MSE, R-squared	Accuracy, precision, recall, F1

Knowing when to use regression or classification depends on the nature of the problem and the target variable. By understanding these distinctions, you'll be better equipped to select the right approach for a given task, laying a strong foundation for effective supervised learning models.

Difference Between Target Variable and Features

In supervised learning, understanding the structure of your data is key to building effective models, and this structure is defined by two primary components: the **target variable** and **features**. The target variable, also called the "dependent variable" or "label," is the value the model aims to predict. Features, also known as "independent variables" or "predictors," are the attributes or inputs used to make predictions. Recognizing the distinction between these two components is essential for constructing reliable supervised learning models, as it defines how your algorithm learns and interprets relationships within the data.

This section will explore the difference between target variables and features, why they matter, and how to effectively prepare and select them for machine learning tasks.

What is the Target Variable?

The target variable is the output or response that we want the model to predict or classify. In a regression task, the target variable is continuous, such as a predicted price or temperature, while in classification, it's categorical, indicating a label or class, such as "spam" or "not spam." The type of target variable fundamentally influences the choice of model and evaluation metrics.

Examples of Target Variables in Real-World Problems

- **House Price Prediction**: The target variable is the price of the house, which is continuous and suitable for regression.
- **Email Spam Detection**: The target variable is whether an email is spam or not, a binary categorical outcome appropriate for classification.
- **Credit Risk Assessment**: The target variable could be a score or a category like "low risk" or "high risk," indicating the creditworthiness of a borrower.

Why the Target Variable Matters

The target variable serves as the model's learning objective, guiding it to identify patterns and relationships that best predict the outcome. In supervised learning, the model uses labeled examples—where both input features and the target variable are known—to learn these relationships.

When defining a machine learning problem, careful consideration of the target variable is crucial for the following reasons:

1. **Choice of Algorithm**: Regression models are used for continuous targets, while classification models handle categorical targets. A mismatch can lead to inappropriate model choice and poor results.
2. **Evaluation Metrics**: The nature of the target variable influences how we evaluate model performance. For example, continuous targets are evaluated with metrics like Mean Absolute Error (MAE) and R-squared, while categorical targets are assessed using accuracy, precision, recall, and F1-score.
3. **Data Preparation**: The target variable's type affects preprocessing steps, such as encoding categorical targets or scaling continuous ones. Accurate handling ensures the model can correctly interpret and predict the target.

What are Features?

Features are the attributes or variables used by the model to learn and make predictions. They represent the information we think is relevant for predicting the target variable. In a dataset, features are the columns (except the target column), and they form the input data the model trains on.

Features can be numerical (e.g., age, income, height), categorical (e.g., color, city, job title), or even textual (e.g., a product review or news article). Some models, such as decision trees, can handle a mix of feature types directly, while others require numerical representation, meaning categorical features may need to be transformed into numerical values before use.

Examples of Features in Real-World Problems

- **Predicting House Prices**: Features might include the number of bedrooms, square footage, location, and age of the house.
- **Email Spam Detection**: Features could include the presence of certain words, email length, and the sender's domain.
- **Predicting Disease Outcome**: Features could consist of age, gender, medical history, and lab test results.

Each feature contributes differently to the prediction task, with some being more relevant than others. For instance, in predicting house prices, square footage might be more significant than the number of bathrooms. Feature selection, which involves identifying the most relevant features, plays a crucial role in improving model accuracy and interpretability.

Why Features Matter

Features provide the informational basis for the model's predictions. Selecting the right features is often as important as choosing

Chapter 8: Introduction to Regression and Classification

the model itself, as irrelevant or noisy features can lead to inaccurate predictions and reduced model performance. Here's why feature selection and engineering are crucial:

1. **Model Accuracy**: Relevant features allow the model to capture the necessary patterns, while irrelevant features may introduce noise, hindering the model's ability to learn effectively.
2. **Model Complexity**: Too many features can increase model complexity, potentially leading to overfitting (where the model memorizes rather than generalizes from the training data). Reducing the feature set to the most essential ones improves model interpretability and efficiency.
3. **Training Time**: Fewer features mean reduced computational cost, making it faster to train and deploy the model, especially important for large datasets and real-time applications.
4. **Interpretability**: Models with well-selected features are easier to understand, which is valuable when explaining model decisions to stakeholders or ensuring transparency in high-stakes applications like healthcare or finance.

Preparing Target Variables and Features for Model Training

Once the target variable and features are identified, the next step is to prepare them for model training. Preparation involves several steps, each critical for ensuring data is clean, accurate, and in a format that the model can interpret effectively.

1. Data Cleaning

Data often comes with inconsistencies, missing values, or outliers, which can hinder model performance. Cleaning ensures that both the target variable and features are in a consistent format and free of errors.

- **Missing Values**: Missing data in the target variable can be problematic as it removes critical information. In features, missing values can be filled with statistical estimates (like the mean or median) or by using more sophisticated imputation methods.
- **Outliers**: Outliers in the target variable or features can distort the model's learning process. Identifying and handling outliers is essential, particularly in regression tasks where extreme values in the target variable can significantly affect model predictions.

2. Encoding Categorical Features

Many algorithms require numerical input, meaning categorical features need transformation. Techniques include:

- **One-Hot Encoding**: Converts categorical values into binary variables, with one variable per category. Useful for nominal features without inherent ordering (e.g., color or city).
- **Label Encoding**: Assigns an integer to each category, effective for ordinal features with an inherent order (e.g., "small," "medium," "large").

3. Feature Scaling

Features with different scales can create biases in model learning, particularly in algorithms sensitive to distance, like K-Nearest Neighbors or linear regression. Feature scaling techniques, such as **standardization** (scaling to a standard deviation of 1) or **normalization** (scaling to a [0, 1] range), ensure that features contribute equally to the model.

4. Feature Engineering

Feature engineering involves creating new features from existing ones or transforming them to provide better insights. Examples include:

- **Combining Features**: Creating a new feature by combining existing ones, like adding age and income to represent a more holistic socioeconomic factor.
- **Polynomial Features**: Raising features to powers (e.g., x^2) to capture non-linear relationships in regression tasks.

Relationship Between Target Variables and Features

In supervised learning, features and the target variable are interconnected through a model, which learns how input features impact the output or target variable. For example, in predicting house prices, features like the number of rooms, location, and lot size collectively influence the target variable, the price. The model learns this relationship by identifying patterns within labeled examples, allowing it to make accurate predictions on new data.

It's crucial to avoid **data leakage**, where information from the target variable inadvertently appears in the features. Data leakage can lead to artificially high accuracy during training, but the model fails on unseen data because it "learned" from information it wouldn't have in real-world scenarios.

Conclusion

The distinction between target variables and features is fundamental in supervised learning. The target variable defines what we want to predict, guiding the model's purpose, while features provide the input data needed to make predictions. Careful selection, preparation, and transformation of features and the target variable lay the groundwork for model success, enabling accurate, interpretable, and efficient machine learning models.

By mastering this distinction, data scientists ensure that their models are built on a solid foundation, improving the chances of accurate predictions and insightful results.

Supervised Learning Algorithms: Linear Regression and Logistic Regression

Supervised learning enables us to build predictive models based on labeled data, and two of the most widely used algorithms in this approach are **Linear Regression** and **Logistic Regression**. Each algorithm has its strengths and limitations, making it suited for different types of problems: Linear Regression is used for predicting continuous values, while Logistic Regression is ideal for binary classification tasks. Together, they serve as foundational tools for understanding supervised learning, providing the groundwork for more advanced algorithms.

In this section, we'll delve into the mechanics, applications, and practical implementation of both Linear and Logistic Regression. We'll explore their mathematical foundations, their typical use cases, and a step-by-step approach to building these models in Python.

1. Linear Regression

Linear Regression is one of the simplest yet most powerful algorithms in supervised learning, especially for tasks that require predicting continuous variables. At its core, Linear Regression tries to find a straight-line relationship between input features (independent variables) and the target variable (dependent variable).

Chapter 8: Introduction to Regression and Classification

How Linear Regression Works

The goal of Linear Regression is to model the relationship between one or more features (predictors) and a continuous target variable by fitting a line that best represents the data. This line is defined by the linear equation:

$$[y = b\_0 + b\_1 \cdot x\_1 + b\_2 \cdot x\_2 + ... + b\_n \cdot x\_n]$$

Here:

- (y) is the predicted value of the target variable.
- (b_0) is the intercept (the point where the line crosses the y-axis).
- $(b\_1, b\_2, ..., b\_n)$ are the coefficients, or slopes, that represent the change in (y) for a unit increase in each respective feature $(x\_1, x\_2, ..., x\_n)$.

The coefficients (b_i) are learned from the data through a process called **optimization**, typically by minimizing a cost function known as **Mean Squared Error (MSE)**. MSE calculates the average of the squared differences between the actual target values and the predicted values, aiming to find the line that minimizes this error.

Types of Linear Regression

- **Simple Linear Regression**: Involves only one predictor variable and the target variable, resulting in a straightforward linear equation with one slope and an intercept.
- **Multiple Linear Regression**: Involves more than one predictor variable, allowing for a more complex model that accounts for multiple factors affecting the target variable.

Applications of Linear Regression

Linear Regression is widely used in scenarios where the objective is to predict a numerical value based on observed trends and relationships. Common applications include:

- **Real Estate**: Predicting property prices based on features like square footage, number of rooms, and location.
- **Sales Forecasting**: Estimating future sales based on historical sales data and seasonal trends.
- **Healthcare**: Predicting patient recovery times or treatment costs based on patient characteristics.

Implementing Linear Regression in Python

Let's walk through a Python example to illustrate how Linear Regression can be implemented using `scikit-learn`:

```
from sklearn.model_selection import
    train_test_split
from sklearn.linear_model import LinearRegression
from sklearn.metrics import mean_squared_error

# Example data: [number of rooms, age of house,
# square footage]
X = [
    [3, 20, 1500],
    [2, 30, 800],
    [4, 10, 2000],
    [3, 15, 1200],
    [5, 8, 2500],
    [3, 25, 1400],
    [4, 5, 2200],
    [2, 40, 700],
    [5, 12, 2600],
```

```
18     [4, 10, 1800]
19  ]
20
21  # Target values (house prices in thousands of
22  # dollars)
23  y = [250, 150, 300, 200, 400, 230, 310, 130, 390,
24     320]
25
26  # Split data into training and testing sets
27  X_train, X_test, y_train, y_test =
28     train_test_split(X, y, test_size=0.2,
29     random_state=42)
30
31  # Initialize and train the model
32  model = LinearRegression()
33  model.fit(X_train, y_train)
34
35  # Make predictions and calculate the error
36  predictions = model.predict(X_test)
37  mse = mean_squared_error(y_test, predictions)
38  print("Mean Squared Error:", mse)
```

This code snippet demonstrates how Linear Regression is used to fit a model, make predictions, and evaluate its accuracy using Mean Squared Error.

---

2. Logistic Regression

Despite its name, Logistic Regression is not used for regression tasks; rather, it is a classification algorithm. Logistic Regression is particularly useful for binary classification, where the goal is to predict one of two possible outcomes (e.g., spam vs. not spam, positive vs. negative).

How Logistic Regression Works

Logistic Regression estimates the probability that a given input belongs to a specific class. It transforms linear predictions into probabilities using the **sigmoid function**, defined as:

$$\sigma(z) = \frac{1}{1 + e^{-z}}$$

where (z) is the output of the linear equation ($b_0 + b_1 \cdot x_1 + b_2 \cdot x_2 + ... + b_n \cdot x_n$).

The sigmoid function maps any real-valued number to a value between 0 and 1, making it ideal for predicting probabilities. If the predicted probability is above a certain threshold (commonly 0.5), the instance is classified as the positive class; otherwise, it's classified as the negative class.

Types of Logistic Regression

- **Binary Logistic Regression**: Used for binary classification tasks, where the target variable has two possible classes.
- **Multinomial Logistic Regression**: Handles multi-class classification problems, where the target variable has three or more possible classes.

Applications of Logistic Regression

Logistic Regression is highly effective for tasks where the goal is to classify data into distinct categories. It's commonly used in:

- **Email Classification**: Predicting whether an email is "spam" or "not spam."
- **Healthcare**: Diagnosing diseases based on patient data (e.g., whether a tumor is malignant or benign).
- **Marketing**: Predicting customer churn (i.e., whether a customer will continue to use a service or leave).

Implementing Logistic Regression in Python

To demonstrate Logistic Regression, we'll use an example where we classify samples based on some features using `scikit-learn`:

```
from sklearn.model_selection import
    train_test_split
from sklearn.linear_model import
    LogisticRegression
from sklearn.metrics import accuracy_score,
    confusion_matrix

# Example data: [age, income (in thousands),
# family size]
X = [
    [25, 50, 1],
    [45, 80, 2],
    [30, 60, 0],
    [35, 40, 3],
    [50, 90, 2],
    [23, 30, 0],
    [40, 70, 1],
    [36, 65, 2],
    [52, 85, 3],
    [28, 55, 1]
]
# Target values: 1 for "Will Buy" and 0 for "Will
# Not Buy"
y = [1, 1, 0, 0, 1, 0, 1, 1, 1, 0]

# Split data into training and testing sets
X_train, X_test, y_train, y_test =
    train_test_split(X, y, test_size=0.2,
    random_state=42)

# Initialize and train the model
```

```
32  model = LogisticRegression()
33  model.fit(X_train, y_train)
34
35  # Make predictions and evaluate accuracy
36  predictions = model.predict(X_test)
37  accuracy = accuracy_score(y_test, predictions)
38  conf_matrix = confusion_matrix(y_test,
39      predictions)
40  print("Accuracy:", accuracy)
41  print("Confusion Matrix:\n", conf_matrix)
```

In this example, the Logistic Regression model classifies data points into one of two classes and evaluates accuracy using the **confusion matrix** and **accuracy score**. The confusion matrix provides additional insight by showing the true positives, true negatives, false positives, and false negatives, which is valuable in understanding model performance beyond simple accuracy.

Comparing Linear Regression and Logistic Regression

While both Linear and Logistic Regression are based on similar underlying principles, they serve distinct purposes:

Aspect	Linear Regression	Logistic Regression
Purpose	Predicting continuous values	Classifying binary (or multiclass) outcomes

Chapter 8: Introduction to Regression and Classification

Aspect	Linear Regression	Logistic Regression
Target Variable	Continuous (e.g., price)	Categorical (e.g., spam vs. not spam)
Output	Real numbers	Probability values (0 to 1)
Equation	$(y = b\_0 + b\_1x\_1 + ... + b\_nx\_n)$	$(\sigma(z) = \frac{1}{1 + e^{-z}})$
Common Applications	Sales forecasting, real estate pricing	Email classification, medical diagnoses

The choice between Linear and Logistic Regression depends on the nature of the problem. For continuous predictions, Linear Regression is suitable, while for binary or categorical classifications, Logistic Regression is preferred.

Conclusion

Linear Regression and Logistic Regression are essential supervised learning algorithms, each suited to specific types of problems. Linear Regression is powerful for tasks that involve predicting a continuous target variable, while Logistic Regression is effective for binary classification problems where outputs fall into distinct categories. Mastering these algorithms lays a foundation for understanding and implementing more complex machine learning models, as they provide insights into key principles such as optimization, probability, and error minimization.

In the next section, we'll apply these concepts with a hands-on activity: using Logistic Regression to predict Titanic survival probabilities, reinforcing these techniques in a practical, real-world context.

Here's the hands-on section using the Titanic dataset with Logistic Regression:

Hands-on Activity: Predicting Titanic Survival with Logistic Regression

For this hands-on activity, we'll apply Logistic Regression to the Titanic dataset to predict which passengers survived the tragic event. This dataset provides real-world context and includes several features, such as age, sex, passenger class, and fare, which can influence survival probability. Logistic Regression is ideal for this binary classification task because it allows us to predict probabilities, which aligns well with survival prediction.

Step 1: Loading and Exploring the Data

The Titanic dataset is available in several Python libraries, including seaborn and pandas. For this example, we'll load it using seaborn.

```
import seaborn as sns
import pandas as pd
from sklearn.linear_model import
    LogisticRegression

# Load the Titanic dataset
titanic = sns.load_dataset("titanic")

# Display the first few rows of the dataset to
# understand its structure
print(titanic.head())
```

Chapter 8: Introduction to Regression and Classification

The dataset includes the following columns:

- **Survived**: Target variable (1 = survived, 0 = did not survive).
- **Pclass**: Passenger class (1 = first, 2 = second, 3 = third).
- **Sex**: Gender of the passenger.
- **Age**: Age of the passenger.
- **SibSp**: Number of siblings or spouses aboard.
- **Parch**: Number of parents or children aboard.
- **Fare**: Fare paid by the passenger.
- **Embarked**: Port of embarkation (C = Cherbourg, Q = Queenstown, S = Southampton).

To prepare the data, we'll focus on a few critical features that could impact survival, such as `Pclass`, `Sex`, `Age`, and `Fare`.

Step 2: Data Preparation

Before training the model, we need to preprocess the data:

1. **Handling Missing Values**: Some columns, like `Age`, have missing values. We can fill these with the median age.
2. **Encoding Categorical Variables**: We'll convert categorical features such as `Sex` and `Pclass` into numeric format. For example, we can convert `Sex` to binary (0 for male, 1 for female).
3. **Feature Selection**: We'll use only relevant features for simplicity, selecting `Pclass`, `Sex`, `Age`, and `Fare`.

```
1   # Handle missing values by filling with median
2   # values
3   import seaborn as sns
4   import pandas as pd
5
6   # Load the Titanic dataset
7   titanic = sns.load_dataset("titanic")
8
9   # Display the first few rows of the dataset to
10  # understand its structure
11  print(titanic.head())
12
13  # Handle missing values by filling with median
14  # values
15  titanic['age'] = titanic['age'].fillna
16      (titanic['age'].median())
17  titanic['fare'] = titanic['fare'].fillna
18      (titanic['fare'].median())
19
20  # Encode categorical variables
21  titanic['sex'] = titanic['sex'].map({'male': 0,
22      'female': 1})
23
24  # Select features and target variable
25  X = titanic[['pclass', 'sex', 'age', 'fare']]
26  y = titanic['survived']
```

Step 3: Splitting the Data

To evaluate the model's performance, we split the data into training and testing sets.

```
1   # Split the data into training and testing sets
2   X_train, X_test, y_train, y_test =
3       train_test_split(X, y, test_size=0.2,
4       random_state=42)
```

Step 4: Building and Training the Model

With the data prepared, we can initialize the Logistic Regression model and fit it to the training data.

```
1   # Initialize and train the Logistic Regression
2   # model
3   model = LogisticRegression()
4   model.fit(X_train, y_train)
```

Step 5: Making Predictions and Evaluating the Model

Once trained, we use the model to predict survival on the test set. We then evaluate its performance using accuracy, a confusion matrix, and classification metrics.

```
1    # Import metric libraries
2    from sklearn.metrics import accuracy_score,
3        confusion_matrix, classification_report
4    # Make predictions
5    predictions = model.predict(X_test)
6
7    # Evaluate the model
8    accuracy = accuracy_score(y_test, predictions)
9    conf_matrix = confusion_matrix(y_test,
10       predictions)
11   classification_rep = classification_report(y_test
12       , predictions)
13
```

```
14    print("Accuracy:", accuracy)
15    print("Confusion Matrix:\n", conf_matrix)
16    print("Classification Report:\n",
17        classification_rep)
```

Output Accuracy: 0.8044692737430168 Confusion Matrix: [[90 15] [20 54]] Classification Report:

```
1              precision    recall  f1-score   support
2
3          0       0.82      0.86      0.84       105
4          1       0.78      0.73      0.76        74
5   accuracy                           0.80       179
6   macro avg       0.80      0.79      0.80       179
7   weighted avg    0.80      0.80      0.80       179
```

- **Accuracy**: Measures the percentage of correctly predicted instances.
- **Confusion Matrix**: Shows true positives, true negatives, false positives, and false negatives, giving insight into the types of errors the model makes.
- **Classification Report**: Provides precision, recall, and F1-score for each class, which are particularly useful when analyzing performance in imbalanced datasets.

Interpretation of Results

Let's break down each metric in more detail as it relates to the Titanic dataset results and discuss what each metric reveals about the model's performance.

1. Accuracy

- **Accuracy**: (0.804) or 80.4%

- Accuracy measures the percentage of all correct predictions, including both survivors and non-survivors. An 80.4% accuracy means the model correctly predicted the survival status of passengers 80.4% of the time.
- While accuracy is a good starting point, it doesn't tell us about performance on individual classes, which can be critical in imbalanced datasets (e.g., many more passengers did not survive than those who did).

2. Confusion Matrix

The confusion matrix provides a breakdown of the model's correct and incorrect predictions for each class. Here's the matrix:

$$\begin{bmatrix} 90 & 15 \\ 20 & 54 \end{bmatrix}$$

Each value can be interpreted as follows:

- **True Negatives (TN)**: 90 passengers who did not survive were correctly predicted as non-survivors.
- **False Positives (FP)**: 15 passengers who did not survive were incorrectly predicted as survivors.
- **False Negatives (FN)**: 20 passengers who survived were incorrectly predicted as non-survivors.
- **True Positives (TP)**: 54 passengers who survived were correctly predicted as survivors.

This breakdown helps identify where the model performs well and where it struggles:

- The model correctly predicts the majority of non-survivors, as indicated by the high true negative count (90). However, it misclassifies some non-survivors as survivors (15).
- For the survivors, the model correctly identifies 54 passengers but misses 20 (false negatives). These false negatives suggest the model struggles somewhat with recognizing survivors.

3. Classification Report

The classification report provides metrics such as **precision, recall,** and **F1-score** for each class (survived = 1, did not survive = 0).

Breakdown by Class

1. **Class 0 (Did Not Survive)**

   - **Precision**: 0.82 – Of all the passengers predicted as non-survivors, 82% were correct.
   - **Recall**: 0.86 – Out of all actual non-survivors, the model correctly identified 86%.
   - **F1-Score**: 0.84 – The harmonic mean of precision and recall. A high F1-score here shows a balanced performance between precision and recall, indicating that the model is quite effective in predicting non-survivors accurately.

2. **Class 1 (Survived)**

   - **Precision**: 0.78 – Of all passengers predicted as survivors, 78% were correct.
   - **Recall**: 0.73 – Out of all actual survivors, the model correctly identified 73%.
   - **F1-Score**: 0.76 – This value is slightly lower than for Class 0, suggesting a modest trade-off between precision and recall for survivors. This discrepancy means that, while the model generally performs well, it may occasionally miss some true survivors or incorrectly predict some non-survivors as survivors.

Summary of Insights and Areas for Improvement

- **Balanced Performance**: Overall, the model shows balanced performance with similar F1-scores for both classes, indicating it can generally distinguish between survivors and non-survivors.
- **Non-Survivor Bias**: The model performs slightly better in predicting non-survivors, which could indicate a slight bias toward the more common class (non-survivors), given that the majority of Titanic passengers did not survive.
- **Missed Survivors (False Negatives)**: The model has a tendency to miss survivors (20 false negatives), which could be improved by refining feature selection or model tuning. Reducing false negatives would improve the model's recall for survivors, making it more sensitive to identifying passengers likely to survive.

In summary, while the model's performance is solid, slight improvements could focus on reducing false negatives for survivors, which might be achieved through hyperparameter tuning or exploring additional features that better distinguish between the survival probabilities of passengers.

Conclusion

This hands-on activity demonstrates the application of Logistic Regression to a real-world dataset, showcasing each step in the machine learning workflow: data preparation, model training, prediction, and evaluation. The Titanic dataset provides a clear example of binary classification, and this exercise reinforces the techniques for implementing and assessing a Logistic Regression model.

In the following chapters, we'll expand on these foundational techniques, exploring other supervised learning algorithms that address more complex and diverse data challenges.

Reflective Questions for Chapter 8 on Regression and Classification:

1. **Fundamental Understanding:**

   - How do you differentiate between regression and classification tasks in supervised learning?

2. **Practical Applications:**

   - Can you think of an industry problem that could benefit from regression analysis? What about classification?

3. **Model Selection:**

   - How would you decide whether to use a linear regression or logistic regression model for a new dataset?

4. **Real-World Impact:**

   - Discuss how regression or classification models have impacted a real-world scenario positively and negatively.

5. **Algorithm Suitability:**

   - For a dataset with multiple categorical outputs, what supervised learning method would you choose and why?

6. **Performance Metrics:**

   - Why is it important to choose the right metrics for evaluating regression and classification models? Can you think of a scenario where using the wrong metric could lead to misleading model evaluations?

Chapter 8: Introduction to Regression and Classification

7. **Feature Importance:**

   - How do the features used in a model affect its accuracy and reliability? Can you provide an example where feature selection dramatically changed a model's performance?

8. **Overfitting Concerns:**

   - What steps would you take to avoid overfitting in a regression or classification model?

9. **Evolution of Algorithms:**

   - How have regression and classification algorithms evolved over time? What are some of the latest developments in this area?

10. **Ethical Implications:**

    - What are the ethical considerations when implementing regression or classification models, especially in sensitive areas like healthcare or criminal justice?

Python Challenges for Understanding Regression and Classification:

1. **Build a Simple Linear Regression Model:**

   - Create a simple linear regression model using synthetic data to predict a target variable based on one feature.

```
1    import numpy as np
2    from sklearn.linear_model import LinearRegression
3    import matplotlib.pyplot as plt
4
5    # Generate synthetic data
6    X = 2 * np.random.rand(100, 1)
7    y = 4 + 3 * X + np.random.randn(100, 1)
8
9    # Build the Linear Regression model
10   model = LinearRegression()
11   model.fit(X, y)
12   y_pred = model.predict(X)
13
14   # Plot the results
15   plt.scatter(X, y, color='blue')
16   plt.plot(X, y_pred, color='red')
17   plt.title('Simple Linear Regression')
18   plt.xlabel('Feature')
19   plt.ylabel('Target')
20   plt.show()
```

2. **Classify Binary Data with Logistic Regression:**

   - Use logistic regression to classify synthetic binary data.

```
1    from sklearn.datasets import make_classification
2    from sklearn.model_selection import train_test_split
3    from sklearn.linear_model import LogisticRegression
4    from sklearn.metrics import classification_report
5
6    # Generate synthetic binary classification data
7    X, y = make_classification(n_samples=100, n_features=2, n\
8    _classes=2, n_clusters_per_class=1, random_state=42)
9
10   # Split the data
```

Chapter 8: Introduction to Regression and Classification

```
11    X_train, X_test, y_train, y_test = train_test_split(X, y,\
12      test_size=0.2, random_state=42)
13
14    # Build and train the Logistic Regression model
15    # <Create the Logistic Regression model and train here>
16    # < figure out the prediction (y_pred) by fitting the tra\
17    ining data on the model>
18    print(classification_report(y_test, y_pred))
```

3. **Visualize Decision Boundaries:**

- Visualize the decision boundaries of a logistic regression model on synthetic data.

```
1     import numpy as np
2     import matplotlib.pyplot as plt
3     from sklearn.linear_model import LogisticRegression
4     from sklearn.datasets import make_classification
5
6     # Generate synthetic data
7     X, y = make_classification(n_samples=1000, n_features=2, \
8     n_informative=2, n_redundant=0, n_repeated=0, n_classes=2
9     , n_clusters_per_class=1, random_state=42)
10
11    # Train a logistic regression model
12    # <Create and fit the logistic regression model here
13    # (fit to the X and y data)>
14
15    # Create a mesh to plot in
16    x_min, x_max = X[:, 0].min() - 1, X[:, 0].max() + 1
17    y_min, y_max = X[:, 1].min() - 1, X[:, 1].max() + 1
18    xx, yy = np.meshgrid(np.arange(x_min, x_max, 0.02), np.ar\
19    ange(y_min, y_max, 0.02))
20
21    # Predict each point on the mesh
```

```
22   Z = model.predict(np.c_[xx.ravel(), yy.ravel()])
23   Z = Z.reshape(xx.shape)
24
25   # Put the result into a color plot
26   plt.contourf(xx, yy, Z, alpha=0.8)
27   plt.scatter(X[:, 0], X[:, 1], c=y, edgecolors='g')
28   plt.title('Logistic Regression Decision Boundaries')
29   plt.xlabel('Feature 1')
30   plt.ylabel('Feature 2')
31   plt.show()
```

Chapter 9: Unsupervised Learning Basics

Introduction to Clustering and Dimensionality Reduction

In machine learning, we commonly encounter scenarios where we don't have labeled data to guide the model during training. Unlike supervised learning, where each data point has a corresponding label (target variable), unsupervised learning algorithms work with unlabeled data. These models aim to discover underlying patterns or groupings within the data based solely on its structure. This is particularly useful in exploratory data analysis, market segmentation, and anomaly detection, among other applications.

Two central techniques in unsupervised learning are **clustering** and **dimensionality reduction**.

Clustering

Clustering is the process of dividing a dataset into distinct groups or clusters based on feature similarity. It's a foundational method in unsupervised learning that reveals groupings without any prior knowledge of classes or categories. The goal is to identify data points that share similar features and group them together in clusters, so that points within each cluster are more similar to each other than to those in other clusters.

Popular clustering algorithms include:

- **K-Means Clustering**: A straightforward and widely used clustering algorithm that partitions data points into a predefined number of clusters, (k), based on feature similarity.
- **Hierarchical Clustering**: This method builds a tree of clusters by iteratively merging or splitting groups, forming a hierarchy.
- **DBSCAN (Density-Based Spatial Clustering of Applications with Noise)**: This approach groups data based on density, effectively identifying clusters of varying shapes and sizes.

Each clustering method has its own strengths and is suited to specific data types and clustering goals. For example, K-Means is efficient and easy to implement but may struggle with irregularly shaped clusters. Hierarchical clustering provides a visual dendrogram but can be computationally intense for large datasets. Meanwhile, DBSCAN excels at detecting noise and irregular shapes but requires carefully selected parameters to be effective.

_____

Dimensionality Reduction

In many machine learning problems, data comes with a high number of features, making it challenging to process, visualize, and analyze. **Dimensionality reduction** addresses this by transforming data into a lower-dimensional space while preserving its essential structure.

Two widely used dimensionality reduction techniques are:

- **Principal Component Analysis (PCA)**: PCA reduces dimensions by finding new axes (principal components) that capture the most variance in the data. By projecting data onto these new axes, PCA reduces the feature space while retaining the information that explains the greatest variance, aiding in visualization and model performance.
- **t-Distributed Stochastic Neighbor Embedding (t-SNE)**: t-SNE is a nonlinear dimensionality reduction technique that preserves the local structure of data, making it particularly useful for visualizing high-dimensional datasets in 2D or 3D.

Choosing the right dimensionality reduction technique depends on the data's structure and the objective. PCA works well for linearly separable data and feature reduction for supervised models, while t-SNE shines when visualizing complex, nonlinear data structures but is less suitable for high-performance modeling tasks.

Clustering and dimensionality reduction reveal patterns that might not be apparent in raw data, offering insight for subsequent modeling or further analysis. As we progress through this chapter, we'll explore the core principles behind these techniques, their real-world applications, and hands-on examples to deepen your understanding.

Algorithms: K-Means Clustering and Principal Component Analysis (PCA)

Now that we understand the role of clustering and dimensionality reduction, let's delve into two foundational algorithms: **K-Means Clustering** and **Principal Component Analysis (PCA)**. Each serves a unique purpose—K-Means groups data into clusters based on similarity, while PCA reduces data dimensions to simplify analysis. Together, they provide a powerful toolkit for uncovering hidden patterns and simplifying complex data structures.

K-Means Clustering

K-Means Clustering is one of the most widely used clustering algorithms in unsupervised learning. Its popularity comes from its simplicity and effectiveness in partitioning data into a predefined number of clusters, denoted as **k**. Here's a step-by-step look at how K-Means operates:

1. **Choose the Number of Clusters (k)**: Before starting, you define the number of clusters. This choice can be based on prior knowledge, or you can try multiple values and evaluate results.
2. **Initialize Centroids**: The algorithm randomly selects **k** data points as initial cluster centers, known as centroids.
3. **Assign Data Points to the Nearest Centroid**: For each data point, K-Means calculates its distance from each centroid and assigns it to the closest one. Each centroid now represents a cluster containing all points nearest to it.
4. **Update Centroids**: After assigning data points to clusters, the algorithm calculates new centroids by finding the average position of all points in each cluster.
5. **Repeat Until Convergence**: Steps 3 and 4 are repeated iteratively until cluster assignments stop changing, or until the centroids no longer shift significantly.

Challenges and Considerations

Choosing the right number of clusters is crucial, as too many clusters can lead to overfitting, while too few may overlook meaningful structure. The **elbow method** is a common approach for selecting

k. It involves plotting the sum of squared distances (variance) between data points and their nearest centroid and looking for the "elbow," or point where adding more clusters doesn't significantly reduce the variance.

Principal Component Analysis (PCA)

While K-Means focuses on grouping similar data points, **Principal Component Analysis (PCA)** is a dimensionality reduction technique aimed at simplifying the dataset. PCA is particularly useful when dealing with high-dimensional data, where the number of features can hinder analysis and visualization. Here's how PCA works:

1. **Standardize the Data**: PCA begins by standardizing the data (transforming features to have a mean of 0 and standard deviation of 1) to ensure that each feature contributes equally to the analysis.
2. **Calculate the Covariance Matrix**: The covariance matrix captures relationships between features. Features with high covariance tend to vary together, indicating they contain overlapping information.
3. **Compute Eigenvalues and Eigenvectors**: The covariance matrix is decomposed into **eigenvalues** and **eigenvectors**. Eigenvectors represent the directions of the new feature space (principal components), while eigenvalues indicate the amount of variance captured by each component.
4. **Select Principal Components**: To reduce dimensions, PCA selects a subset of the principal components, retaining only those that capture a significant portion of the data's variance. For instance, retaining components that account for 95% of variance can achieve effective dimensionality reduction without significant information loss.

5. **Transform Data**: The original data is projected onto the selected principal components, creating a reduced feature space that simplifies analysis while retaining core information.

Applications and Limitations
PCA is especially valuable in preprocessing steps, such as visualizing high-dimensional data or reducing features before using other algorithms. However, it's best suited for linearly separable data, as it cannot capture nonlinear relationships. In cases where nonlinear structure is important, other dimensionality reduction methods like t-SNE might be more effective.

By understanding K-Means Clustering and PCA, we gain tools to structure data meaningfully and simplify complex datasets. In the next sections, we'll explore how to apply these algorithms in practice, with hands-on examples that showcase their effectiveness in revealing hidden patterns and reducing dimensionality in data.

Let's delve into a **Case Study** section where we'll examine the Kaggle *Customer Segmentation* dataset using Python to implement K-Means clustering.

Case Study: Kaggle Dataset for Clustering - Customer Segmentation

Clustering plays a pivotal role in segmenting data, especially in areas like customer analysis, where understanding patterns can drive targeted marketing and personalized recommendations. In this case study, we'll use the popular **Customer Segmentation** dataset from Kaggle. This dataset contains various customer attributes, such as age, income, and spending scores, which we'll use to group customers into segments.

This practical example demonstrates how to apply K-Means clustering to uncover customer segments, guiding business decisions like personalized marketing campaigns.

---

1. Loading the Dataset

Start by loading and examining the dataset. If you're using Google Colab or Jupyter, download the dataset from Kaggle and upload it to your working environment.

```
1   import pandas as pd
2
3   # Load the dataset
4   data = pd.read_csv("customer_segmentation.csv")
5
6   # Display the first few rows of the dataset
7   data.head()
```

This dataset typically includes columns like Age, Annual Income (k$), and Spending Score (1-100), each providing insights into different customer attributes. Before clustering, ensure data is clean and normalized to prevent biases in clustering due to differing scales.

---

2. Preprocessing the Data

Data normalization is essential, especially for algorithms sensitive to feature scales. Let's use the StandardScaler to standardize features.

```
1   from sklearn.preprocessing import StandardScaler
2
3   # Select relevant features
4   features = data[['age', 'annual_income',
5       'spending_score']]
6
7   # Standardize the features
8   scaler = StandardScaler()
9   scaled_features = scaler.fit_transform(features)
```

By scaling the data, each feature contributes equally to the clustering algorithm.

---

3. Applying K-Means Clustering

With standardized data, we can apply the K-Means algorithm. First, let's use the **elbow method** to determine the optimal number of clusters.

```
1   import matplotlib.pyplot as plt
2   from sklearn.cluster import KMeans
3
4   # Use the elbow method to find the optimal number
5   # of clusters
6   inertia = []
7   k_range = range(1, 11)
8
9   for k in k_range:
10      kmeans = KMeans(n_clusters=k, random_state=0)
11      kmeans.fit(scaled_features)
12      inertia.append(kmeans.inertia_)
13
14  # Plot the elbow graph
```

```
15  plt.figure(figsize=(8, 5))
16  plt.plot(k_range, inertia, marker='o')
17  plt.xlabel('Number of clusters (k)')
18  plt.ylabel('Inertia')
19  plt.title('Elbow Method for Optimal k')
20  plt.show()
```

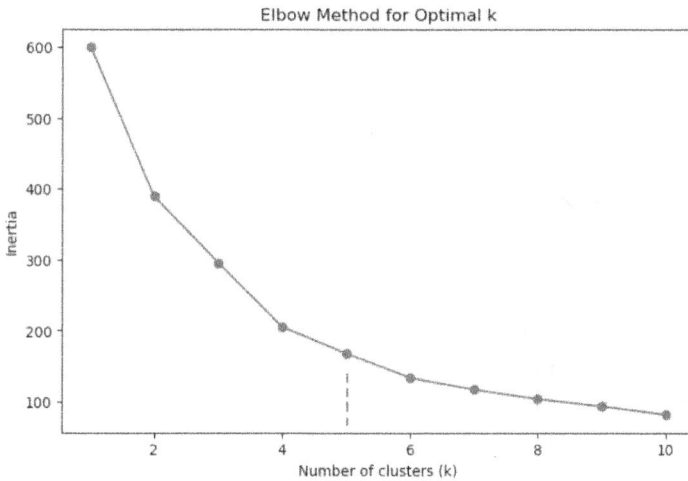

Figure 47. Elbow method for choosing optimal k

The **elbow point** on this plot suggests the optimal number of clusters. Choose this number (e.g., k=5) for clustering.

```
1    # Set the optimal number of clusters
2    optimal_k = 5
3
4    # Initialize and fit the KMeans model
5    kmeans = KMeans(n_clusters=optimal_k,
6        random_state=0)
7    data['Cluster'] = kmeans.fit_predict
8        (scaled_features)
```

Now, each customer is assigned to a cluster, adding a new Cluster column in the dataset.

---

4. Analyzing the Clusters

After clustering, analyze each group to understand customer characteristics. This can be visualized by plotting the clusters in two dimensions.

```
1    import seaborn as sns
2
3    # Plot the clusters
4    plt.figure(figsize=(10, 7))
5    sns.scatterplot(x=data['annual_income'], y=data[
6        'spending_score'],
7                    hue=data['Cluster'],
8        palette='viridis', s=60)
9    plt.title('Customer Segmentation using K-Means
10        Clustering')
11   plt.xlabel('Annual Income (k$)')
12   plt.ylabel('Spending Score (1-100)')
13   plt.legend(title='Cluster')
14   plt.show()
```

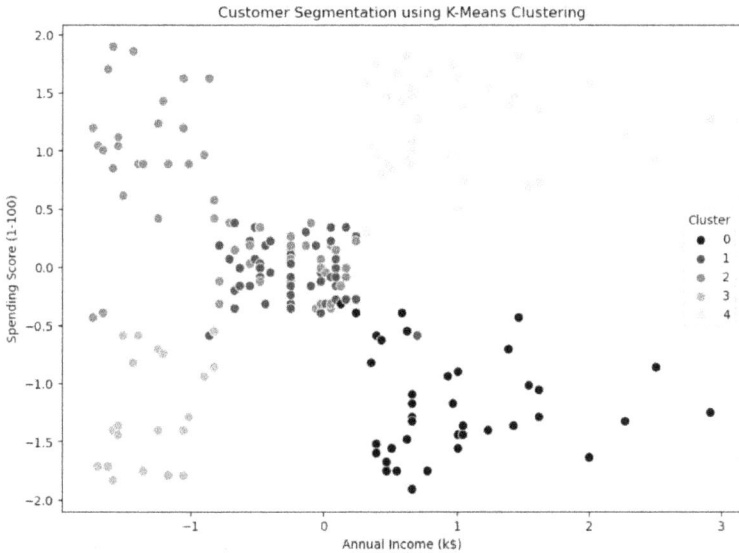

Figure 48. K-Mean Clustering

Clusters reveal customer groups with similar income and spending patterns, allowing businesses to develop tailored strategies for each group. For example, high-income, high-spending customers may be prime candidates for premium products or loyalty programs.

5. Insights and Next Steps

The K-Means clustering approach has helped us identify key customer segments within the data. These segments can inform marketing strategies, customer service approaches, and product offerings. Moving forward, consider experimenting with different clustering algorithms (e.g., **Hierarchical Clustering** or **DBSCAN**) or dimensionality reduction techniques like PCA to further refine insights.

To incorporate PCA as a follow-up to the initial K-Means clustering, we'll demonstrate how PCA can simplify high-dimensional data after performing K-Means. This approach lets us compare clustering performance on the original features versus PCA-reduced features, ultimately showing how PCA aids in visualization and potentially improves clustering efficiency.

Here's how to structure this additional PCA section:

Case Study: Adding PCA to Improve Visualization of K-Means Clustering

After performing K-Means clustering on the original dataset, let's apply **Principal Component Analysis (PCA)** to reduce our features to two principal components. This will help us visualize the clusters in a simplified 2D space, retaining key data patterns but making it easier to interpret.

1. Applying PCA to Reduce Dimensionality

We'll apply PCA to reduce our three original features (e.g., `age`, `annual_income`, `spending_score`) to two principal components. This reduction simplifies visualization while preserving the majority of the data's variability.

```
1  from sklearn.decomposition import PCA
2
3  # Apply PCA to reduce features to 2 dimensions
4  pca = PCA(n_components=2)
5  pca_features = pca.fit_transform(scaled_features)
6
7  # Add the PCA-transformed components to our
8  # DataFrame
9  data['PCA1'] = pca_features[:, 0]
10  data['PCA2'] = pca_features[:, 1]
```

2. Plotting the K-Means Clusters in PCA-Reduced Dimensions

Now that we've reduced the data to two dimensions, let's plot the clusters to see if the customer segments identified by K-Means remain clear and distinct.

```
1  import seaborn as sns
2  import matplotlib.pyplot as plt
3
4  # Plot the clusters in the PCA-reduced 2D space
5  plt.figure(figsize=(10, 7))
6  sns.scatterplot(x='PCA1', y='PCA2',
7      hue=data['Cluster'], palette='viridis', s=60,
8      data=data)
9  plt.title('Customer Segmentation using PCA
10      -Reduced Features and K-Means Clustering')
11  plt.xlabel('Principal Component 1')
12  plt.ylabel('Principal Component 2')
13  plt.legend(title='Cluster')
14  plt.show()
```

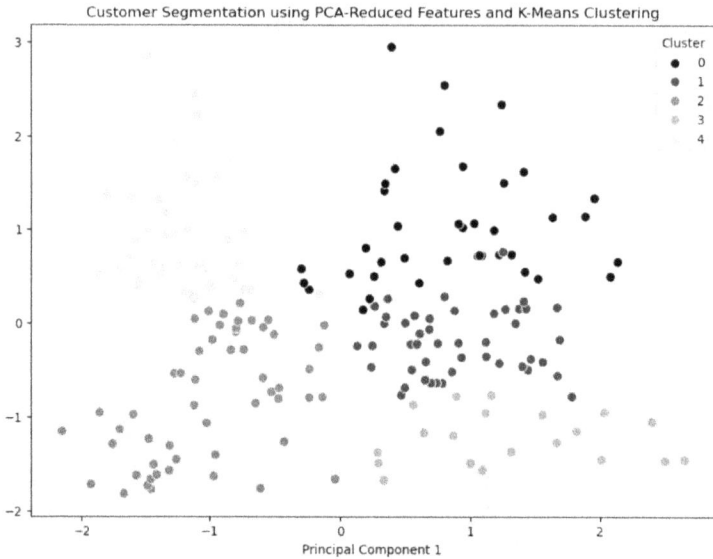

Figure 49. Clustering using PCA to reduce features

This plot provides a visual comparison, showing how clustering results translate into a 2D space. While K-Means identified clusters on the original features, PCA helps reveal these clusters' separations more clearly by reducing noise and overlapping features.

---

3. Interpreting the Results with PCA and K-Means

By applying PCA after K-Means clustering, we gain insight into the cluster structure in a simplified view:

- PCA-reduced features help clarify relationships and make clusters more distinct.
- This visualization can help identify key clusters (e.g., high spenders or budget-conscious customers) for targeted marketing or further analysis.

Hands-on Activity: Perform Clustering on the *Wisconsin Cancer* Dataset

In this hands-on activity, we'll apply clustering to the *Wisconsin Cancer* dataset. This dataset contains various features related to tumor characteristics, and clustering can help identify potential groups, which may correlate with benign and malignant types or reveal subgroups for further medical study.

Step 1: Load the Dataset and Data Process

If you're using Google Colab or Jupyter Notebook, download the Wisconsin Cancer dataset, ensuring it's available in your environment.

```python
import pandas as pd
from sklearn.preprocessing import StandardScaler

# Load the dataset
data = pd.read_csv("wisconsin_cancer.csv")

# Select relevant features and diagnosis
features = data[['radius_mean', 'texture_mean',
    'perimeter_mean', 'area_mean']]
diagnosis = data['diagnosis']

# Standardize the features
scaler = StandardScaler()
scaled_features = scaler.fit_transform(features)
```

Step 2: Perform K-Means Clustering

Use the elbow method to find the optimal number of clusters for K-Means. The elbow method helps us select the optimal number of

clusters by observing the point where adding more clusters doesn't
substantially reduce inertia.

```python
import matplotlib.pyplot as plt
from sklearn.cluster import KMeans
import warnings

warnings.filterwarnings("ignore")

# Elbow method to determine the optimal number of
# clusters
inertia = []
k_range = range(1, 11)

for k in k_range:
    kmeans = KMeans(n_clusters=k, random_state=0)
    kmeans.fit(scaled_features)
    inertia.append(kmeans.inertia_)

# Plot the elbow graph
plt.figure(figsize=(8, 5))
plt.plot(k_range, inertia, marker='o')
plt.xlabel('Number of clusters (k)')
plt.ylabel('Inertia')
plt.title('Elbow Method for Optimal k')
plt.show()
```

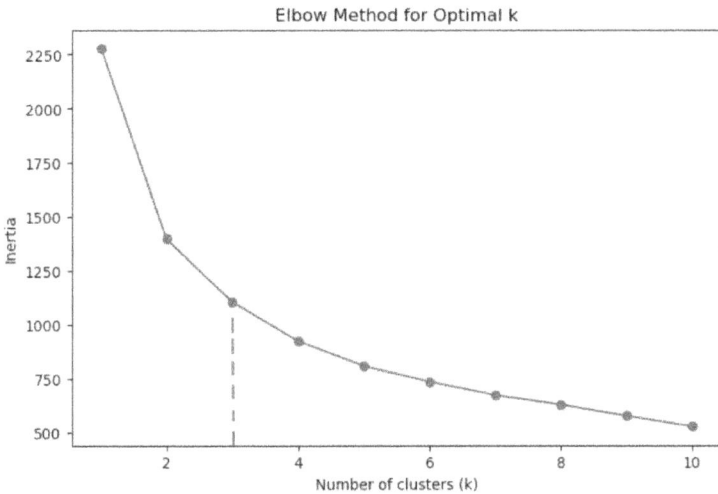

Figure 50. Kmean Fit and Elbow

Step 3: Apply K-Means Clustering

Using the optimal value for k based on the elbow plot, we can now apply K-Means to group the tumor data.

```
# Set the optimal number of clusters (e.g., k=3)

optimal_k = 3
kmeans = KMeans(n_clusters=optimal_k,
    random_state=0)
data['Cluster'] = kmeans.fit_predict
    (scaled_features)
```

Step 4: Optional PCA for Dimensionality Reduction and Visualization

For ease of visualization, we'll use PCA to reduce the data to two principal components, which allows us to plot clusters in a 2D space.

```
1   from sklearn.decomposition import PCA
2   import seaborn as sns
3
4   # Apply PCA to reduce to 2 dimensions
5   pca = PCA(n_components=2)
6   pca_features = pca.fit_transform(scaled_features)
7   data['PCA1'] = pca_features[:, 0]
8   data['PCA2'] = pca_features[:, 1]
9
10  # Plot clusters with PCA components and overlay
11  # diagnosis
12  plt.figure(figsize=(10, 7))
13  sns.scatterplot(x='PCA1', y='PCA2',
14      hue=data['diagnosis'], style=data['Cluster'],
15                  palette={'B': 'blue', 'M': 'red'}
16      , s=60, data=data)
17  plt.title('Wisconsin Cancer Clustering with
18      Diagnosis Labels and PCA')
19  plt.xlabel('Principal Component 1')
20  plt.ylabel('Principal Component 2')
21  plt.legend(title='Diagnosis')
22  plt.show()
```

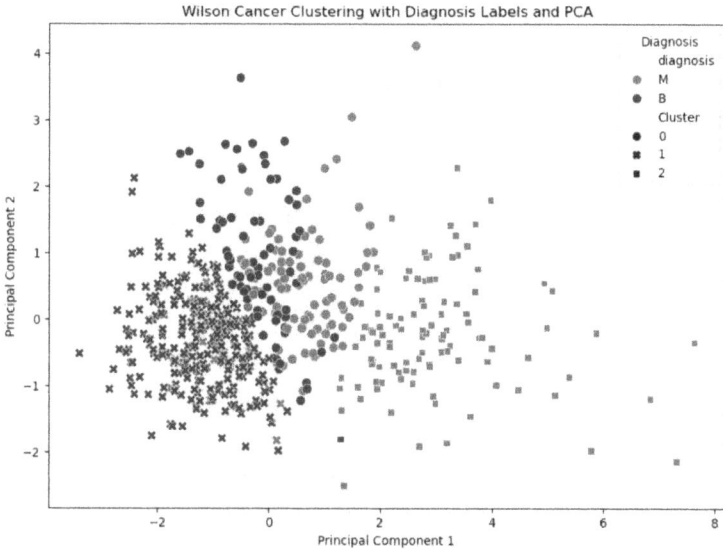

Figure 51. Wisconsin Cancer Clustering with PCA and K-Means

Step 5: Interpreting Cluster Insights

With this enhanced plot, students can observe:

Cluster Patterns: Clusters that predominantly contain points labeled as "M" suggest that the algorithm has grouped malignant cases based on similar tumor characteristics. **Overlap Analysis:** Any areas with both "B" and "M" labels may indicate tumor characteristics shared between benign and malignant types, potentially highlighting challenging cases. **Cluster Purity:** If clusters align well with "B" (benign) and "M" (malignant) labels, it suggests that the selected features provide strong separability for malignancy prediction.

Summary

This activity demonstrates how clustering can reveal significant patterns in medical data, such as potential benign and malignant groupings, enhancing understanding of tumor characteristics. By comparing K-Means clusters with actual diagnoses, students gain practical insights into how clustering supports exploratory data analysis in healthcare.

Reflective Questions for Chapter 9 (K-Means and PCA)

1. **Understanding Unsupervised Learning**: How does unsupervised learning differ from supervised learning in terms of data requirements and outcomes? Why might unsupervised learning be more challenging?

2. **Clustering Techniques Comparison**: What are the key differences between K-Means, Hierarchical, and DBSCAN clustering? In what scenarios might one be preferred over the others?

3. **Importance of Feature Scaling**: Why is feature scaling important in clustering algorithms like K-Means? What might happen if you don't scale your features?

4. **Choosing 'k' in K-Means**: Discuss how the elbow method helps in determining the optimal number of clusters in K-Means. Are there any limitations to this method?

5. **Dimensionality Reduction Outcomes**: Explain how PCA helps in dimensionality reduction. What are the implications of reducing dimensions on data analysis?

6. **Applications of Clustering and PCA**: Can you think of a real-world problem where clustering and PCA might be applied together? How would these techniques help solve the problem?

7. **Interpreting Clustering Results**: After performing clustering, how can the results be used in decision-making processes? Provide an example related to customer segmentation.

8. **Limitations of PCA**: Discuss the limitations of PCA when dealing with nonlinear data. What alternative techniques could be used in such cases?

Python Challenges for Chapter 9

Challenge 1: Explore Clusters with K-Means

```python
# Task: Use scikit-learn to apply K-Means
# clustering to a simple dataset and visualize
# the results.
import matplotlib.pyplot as plt
from sklearn.datasets import make_blobs
from sklearn.cluster import KMeans

# Generate synthetic data
X, y = make_blobs(n_samples=150, centers=3,
    random_state=6)

# Apply K-Means clustering
# <Add your code here>
# Create a model supporting 3 clusters
# and predict the labels of these clusters

# Plotting the clusters
plt.scatter(X[:, 0], X[:, 1], c=labels,
    cmap='viridis')
centers = kmeans.cluster_centers_
plt.scatter(centers[:, 0], centers[:, 1], c='red'
    , s=100, alpha=0.75)  # Mark the centroids
plt.title('K-Means Clustering')
plt.xlabel('Feature 1')
plt.ylabel('Feature 2')
plt.show()
```

Challenge 2: Visualize High-Dimensional Data Using PCA

```
1   # Task: Use PCA from scikit-learn to reduce the
2   # dimensionality of a dataset and plot the
3   # results.
4   from sklearn.decomposition import PCA
5   from sklearn.datasets import make_blobs
6   import matplotlib.pyplot as plt
7
8   # Generate synthetic data with more features
9   X, _ = make_blobs(n_samples=150, centers=3,
10      n_features=4, random_state=42)
11
12  # Apply PCA to reduce dimensions to 2
13  # <Add your code here>
14  # Create the PCA model with 2 dimensions
15  # and fit the data to transform it to only 2
16  # dimensions.
17
18  # Plotting the reduced data
19  plt.scatter(X_pca[:, 0], X_pca[:, 1])
20  plt.title('PCA Dimensionality Reduction')
21  plt.xlabel('Principal Component 1')
22  plt.ylabel('Principal Component 2')
23  plt.show()
```

These tasks are designed to be straightforward, allowing students to focus on understanding the algorithms' effects without delving deeply into the math or more complex aspects of implementation. They also provide visual feedback that helps students see the impact of these algorithms on the data, enhancing their learning experience.

Chapter 10: Decision Trees and Random Forests

As we progress in machine learning, we encounter more sophisticated techniques that allow us to handle complex data, improve model accuracy, and make more robust predictions. Advanced machine learning techniques like decision trees, ensemble methods, and support vector machines can capture nuanced relationships within data and often outperform simpler models on complex datasets. This unit explores these advanced approaches, starting with **Decision Trees** and **Random Forests**, which form the backbone of many modern predictive models.

Introduction to Decision Trees

Decision Trees are intuitive and powerful models used for both classification and regression tasks. They work by splitting data into smaller subsets based on feature values, creating a branching structure that resembles a tree. Each "node" in a decision tree represents a decision based on a feature, and each "leaf" represents a final output or prediction.

One of the greatest advantages of decision trees is their interpretability. Unlike many other machine learning models, decision trees can be visualized and easily understood. This makes them valuable tools when transparency in decision-making is essential, such as in finance, healthcare, or any field where model predictions need to be interpretable.

Here's how decision trees work in a nutshell:

1. **Splitting the Data**: Starting from the root, a decision tree splits data at each node by selecting a feature and a threshold that best separate the target variable. This process continues at each node, creating branches that direct data points down the tree based on their feature values.

2. **Purity of Nodes**: The objective is to make each node as "pure" as possible, meaning that each leaf node ideally contains data points with the same target class (for classification) or similar values (for regression). Different algorithms, such as **Gini Impurity** or **Entropy**, measure the "purity" of nodes and guide the tree in choosing optimal splits.

3. **Recursive Splitting and Tree Depth**: The tree continues splitting until it either reaches a maximum depth, has pure nodes, or meets a criterion for stopping. Trees that grow too deep may "overfit" the data, capturing noise and making predictions that don't generalize well. Controlling tree depth or minimum node sizes can prevent overfitting.

4. **Prediction**: Once the tree is built, making a prediction is straightforward. The model follows the decision path based on input features, leading to a leaf node where the final prediction is made.

In the next section, we'll discuss the algorithms behind choosing optimal splits, including **Gini Impurity** and **Entropy**, as well as techniques to control tree growth and avoid overfitting. This sets the foundation for understanding **Random Forests** and other ensemble methods that build on decision trees to create more robust models.

Advantages and Drawbacks of Decision Trees

Decision trees are popular due to their simplicity and interpretability, but they also have limitations that can impact their performance on certain tasks. Let's explore the main strengths and weaknesses of decision trees to understand when they're most suitable and what to consider before using them.

Advantages of Decision Trees

1. **Interpretability**: Decision trees are highly interpretable, allowing us to visualize and understand how the model arrives at its predictions. Each split is based on a simple rule, making decision trees particularly valuable in fields where transparency is essential.

2. **No Requirement for Feature Scaling**: Unlike models such as linear regression or k-nearest neighbors, decision trees don't require feature scaling or normalization. Each split is based on the actual values of the features, making preprocessing simpler.

3. **Handles Both Numerical and Categorical Data**: Decision trees can handle both types of data, enabling flexibility in data selection and preprocessing. Many tree-based implementations automatically handle categorical variables by creating binary splits.

4. **Nonlinear Relationships**: Decision trees can capture nonlinear relationships because they make decisions based on feature splits rather than assuming any linear relationship between features and the target. This adaptability makes them effective on complex datasets where linear models may struggle.

5. **Feature Importance**: Decision trees provide insights into feature importance by measuring the impact of each feature on the target variable, which can be valuable for feature selection and model interpretation.

Drawbacks of Decision Trees

1. **Overfitting**: Decision trees are prone to overfitting, especially when they grow too deep, capturing noise in the training data and leading to poor generalization on new data. Regularization techniques, such as limiting tree depth or requiring a minimum number of samples per split, are essential to avoid this.

2. **Instability**: Decision trees can be highly sensitive to small changes in the data. Minor adjustments in the dataset can result in different splits and, ultimately, an entirely different tree structure. This instability is due to the greedy nature of the algorithm, which makes each split without considering future splits.

3. **Bias Toward Dominant Features**: Decision trees tend to focus on features with many unique values, which can bias the model if those features are not particularly informative. This characteristic can lead to splits that prioritize high-cardinality features, even when other features may have more predictive power.

4. **High Variance**: Decision trees can have high variance, meaning that their predictions may vary significantly across different subsets of the data. This issue is especially problematic for smaller datasets and is one reason why **ensemble methods** like Random Forests or Gradient Boosting, which combine multiple trees, are often preferred.

5. **Difficulty Capturing Complex Patterns**: Although decision trees are flexible with nonlinear relationships, they may still struggle with capturing complex interactions between features. Ensemble methods like Random Forests or boosting techniques are often necessary to capture these patterns effectively.

Decision trees are valuable models with many practical applications, but they come with limitations, especially in terms of stability and overfitting. Understanding these pros and cons helps in deciding when to use decision trees and when to opt for ensemble techniques, which we'll explore further in this chapter.

Next, we'll dive into overcoming some of the problems with decision trees using an algorithm known as **random forests**.

Random Forests for Better Performance

While decision trees offer interpretability and flexibility, they are prone to overfitting and can be sensitive to small changes in the data. To address these drawbacks, we turn to **Random Forests**, an ensemble technique that builds multiple decision trees and combines their predictions for more accurate and stable results.

Random Forests work by constructing many decision trees, each trained on a different random subset of the data and features. This process introduces randomness, which reduces the model's tendency to overfit and makes predictions more robust. Here's a closer look at how Random Forests improve upon individual decision trees:

---

1. Ensemble of Multiple Trees

A Random Forest consists of hundreds or even thousands of decision trees, known as "weak learners." Each tree is trained on a bootstrapped sample (random subset) of the data, meaning it may contain duplicates and miss some data points. This technique, known as **bagging** (bootstrap aggregating), reduces variance, making the overall model less sensitive to variations in the training set.

2. Random Feature Selection

In addition to sampling data points, Random Forests introduce further randomness by selecting a random subset of features at each split within each tree. This approach prevents the model from relying too heavily on specific, dominant features, encouraging trees to explore different patterns within the data. As a result, the ensemble captures more diverse relationships, enhancing generalization.

3. Aggregation of Predictions

After training, each tree in the Random Forest makes a prediction for a given input. For classification, the final output is determined by a **majority vote** across all trees; for regression, it's the **average of all tree predictions**. This aggregation process reduces the model's variance, providing a more stable and reliable prediction than any single tree could achieve on its own.

4. Improved Generalization and Reduced Overfitting

By averaging the predictions of multiple trees, Random Forests create a strong predictive model that generalizes well to new data. The combined predictions are less likely to be swayed by individual trees that may overfit or capture noise. This improvement is particularly valuable for high-dimensional or noisy datasets where a single decision tree might struggle.

Advantages of Random Forests

- **Robustness**: Random Forests are less sensitive to small changes in the data than individual trees, reducing the high variance typically associated with decision trees.
- **High Accuracy**: The ensemble of trees in a Random Forest often results in more accurate predictions, making it one of the most popular methods in supervised learning.
- **Automatic Feature Importance**: Random Forests provide a measure of feature importance, revealing which features contribute most to the model's predictions. This can assist with feature selection and interpretation.

Drawbacks of Random Forests

- **Reduced Interpretability**: Although each decision tree in a Random Forest is interpretable, the ensemble as a whole is less so, making it challenging to visualize or fully understand the decision-making process.
- **Computationally Intensive**: Training a large number of decision trees requires more computational power and memory than a single tree, which can be a drawback for very large datasets or resource-limited environments.

Random Forests offer a powerful way to harness the strengths of decision trees while mitigating their weaknesses, making them ideal for tasks where high accuracy and generalization are crucial. In the next section, we'll explore how to train and tune Random Forests in Python, leveraging their capabilities to improve model performance in practical applications.

Here's the **Hands-on Activity** section for classifying data with decision trees using the *Heart Disease* dataset:

Hands-on Activity: Use the *Heart Disease* Dataset to Classify Data with Decision Trees

In this hands-on activity, we'll apply a decision tree to classify cases of heart disease. Using the popular *Heart Disease* dataset, which contains patient data and risk factors, we'll build a decision tree model to predict the presence or absence of heart disease. This activity provides a practical introduction to decision trees and demonstrates how they can assist in medical diagnostics by uncovering patterns within health data.

Step 1: Load the Dataset

Start by loading the dataset, typically available on sites like Kaggle or UCI Machine Learning Repository. Ensure the dataset includes columns for various risk factors (such as age, blood pressure, and cholesterol levels) and a target column indicating heart disease presence.

```
1   import pandas as pd
2
3   # Load the dataset
4   data = pd.read_csv("heart_disease.csv")
5
6   # Display the first few rows
7   data.head()
```

Step 2: Data Preprocessing

Before training the model, we need to handle any missing values, select features, and split the dataset into training and testing sets. For simplicity, let's assume the dataset is clean, but always check for any preprocessing needs in real scenarios.

```
1    from sklearn.model_selection import
2        train_test_split
3
4    # Define features (X) and target variable (y)
5    # Assuming 'target' is the column for
6    # heart disease presence
7    X = data.drop(columns=['target'])
8
9    y = data['target']
10
11   # Split the dataset into training and testing sets
12   X_train, X_test, y_train, y_test =
13       train_test_split(X, y, test_size=0.2,
14       random_state=42)
```

Step 3: Train a Decision Tree Classifier

Using Scikit-Learn, we can easily train a decision tree model on the training data. Set parameters like max_depth to control the tree's growth and prevent overfitting.

```
1   from sklearn.tree import DecisionTreeClassifier
2
3   # Initialize the Decision Tree Classifier
4   tree_model = DecisionTreeClassifier(max_depth=4,
5       random_state=42)
6
7   # Train the model
8   tree_model.fit(X_train, y_train)
```

Step 4: Evaluate the Model

After training, evaluate the model on the test set to understand its accuracy and performance. This step includes generating predictions and calculating accuracy.

```
1   from sklearn.metrics import accuracy_score,
2       classification_report
3
4   # Make predictions on the test set
5   y_pred = tree_model.predict(X_test)
6
7   # Calculate accuracy and other metrics
8   accuracy = accuracy_score(y_test, y_pred)
9   print("Accuracy:", accuracy)
10  print("\nClassification Report:\n",
11      classification_report(y_test, y_pred))
```

Output Accuracy: 0.8

Classification Report: precision recall f1-score support

1	0	0.88	0.70	0.78	102
2	1	0.75	0.90	0.82	103
3					
4	accuracy			0.80	205
5	macro avg	0.81	0.80	0.80	205
6	weighted avg	0.81	0.80	0.80	205

Here's an interpretation of the model's performance based on the results:

Overall Accuracy

- **Accuracy**: 0.8 (or 80%)
  This indicates that the model correctly classified 80% of all instances in the dataset. While this is generally a good accuracy score, it's also important to look at precision, recall, and F1-score for a clearer view of how well the model handles each class.

Class-Level Performance

Class 0 (No Heart Disease)

- **Precision**: 0.88
  Precision for class 0 is high, meaning that when the model predicts a patient doesn't have heart disease, it is correct 88% of the time. This suggests the model is effective at minimizing false positives for the "no heart disease" group.
- **Recall**: 0.70
  The recall for class 0 indicates that the model captures 70% of the actual "no heart disease" cases. The lower recall shows some patients without heart disease are being misclassified as having the condition (false negatives).
- **F1-score**: 0.78
  The F1-score, which balances precision and recall, is 0.78.

This suggests good but slightly imbalanced performance, with a tendency to prioritize precision over recall for class 0.

Class 1 (Heart Disease)

- **Precision**: 0.75
  For class 1, the precision is 0.75, meaning that 75% of the times the model predicts "heart disease," it is correct. This is lower than precision for class 0, suggesting more false positives for "heart disease."
- **Recall**: 0.90
  The model correctly identifies 90% of actual heart disease cases, demonstrating high sensitivity. This high recall is valuable in medical applications, where it's often critical to identify as many positive cases as possible.
- **F1-score**: 0.82
  The F1-score for class 1 is slightly higher than for class 0, indicating a stronger balance between precision and recall. This score shows that the model does a good job of identifying heart disease cases without excessively misclassifying non-cases as cases.

Averages

- **Macro Average**: 0.81 (precision), 0.80 (recall), 0.80 (F1-score)
  The macro average calculates metrics independently for each class and then takes the average. Here, the similar scores across metrics indicate that the model maintains a fairly balanced performance between both classes.
- **Weighted Average**: 0.81 (precision), 0.80 (recall), 0.80 (F1-score)

The weighted average considers the support (number of instances) for each class, making it a more realistic reflection of the model's overall performance, especially in imbalanced datasets.

Overall Observation and Recommendation

This model demonstrates solid performance with an overall accuracy of 80% and generally good precision and recall across both classes. It's effective at identifying heart disease cases (high recall for class 1) and reasonably good at predicting no heart disease, though it has slightly more errors in misclassifying healthy cases as diseased. This model would likely be suitable in a clinical setting, where it's generally more critical to avoid missing true cases of heart disease (high recall for class 1).

Step 5: Visualize the Decision Tree

Visualizing the tree helps us understand its decision-making process, which can reveal which features are important for predicting heart disease. Scikit-Learn provides easy integration with libraries like `graphviz` to plot the tree.

```
1   from sklearn.tree import plot_tree
2   import matplotlib.pyplot as plt
3
4   # Plot the decision tree
5   plt.figure(figsize=(20, 10))
6   plot_tree(tree_model, feature_names=X.columns,
7       class_names=['No Heart Disease',
8       'Heart Disease'], filled=True)
9   plt.show()
```

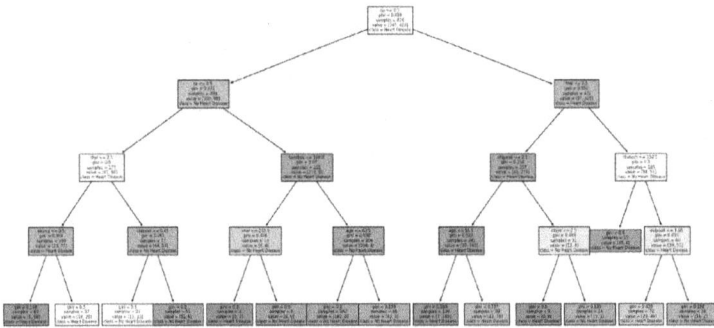

Figure 52. Decision Tree Visualized

Interpreting the Decision Tree

From the visualization, we can observe how the decision tree uses various health indicators to split the data. For example, it might prioritize cholesterol levels or age for certain splits. Each path from the root to a leaf node represents a possible outcome based on feature values, helping us understand how different factors contribute to heart disease risk. Next we'll look how to pull feature importance from our random forest model.

Visualizing Feature Importance in the Decision Tree Model

One of the valuable aspects of decision trees is their ability to highlight which features contribute most to the predictions. This **feature importance** metric reveals how much each feature influences the model's decisions, providing insights into which factors are most relevant for predicting heart disease.

In this section, we use the trained decision tree model to retrieve and visualize feature importance. This helps us understand the underlying data patterns and supports more informed decision-making in medical contexts.

1. **Retrieve Feature Importances**
2. After training the decision tree model, we access `tree_-model.feature_importances_`, which provides a ranking of each feature's importance. This ranking reflects how much each feature contributes to reducing the uncertainty or impurity at each split in the tree. Features with higher values have a stronger influence on the model's predictions.

```
1    feature_importances =
2        tree_model.feature_importances_
3    feature_names = X.columns
```

3. **Create a DataFrame for Visualization**: We organize feature names and their importance values into a DataFrame. Sorting this DataFrame by the importance values allows us to see the most influential features at the top.

```
1    importance_df = pd.DataFrame({'Feature':
2      feature_names, 'Importance':
3      feature_importances})
4    importance_df = importance_df.sort_values
5      (by='Importance', ascending=False)
```

4. **Plot the Feature Importances**: Using a horizontal bar chart, we can clearly see the ranking of features, with the most important ones displayed at the top. This visualization makes it easy to interpret the significance of each feature in predicting heart disease.

```
1    plt.figure(figsize=(10, 8))
2    plt.barh(importance_df['Feature'],
3      importance_df['Importance'], color='skyblue')
4    # Invert y-axis for descending order
5    plt.gca().invert_yaxis()
6    plt.xlabel('Importance')
7    plt.ylabel('Feature')
8    plt.title('Feature Importance in Predicting
9      Heart Disease')
10   plt.show()
```

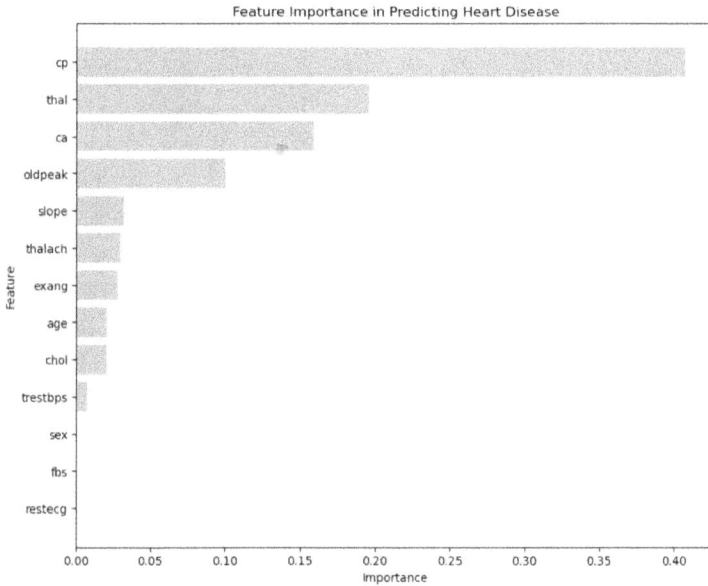

Figure 53. Charting Feature Importance

Interpretation of the Plot:

The plot reveals which features the decision tree model prioritized in making predictions. For example, since chest pain (cp) and thal (thalassemia) appear at the top, it indicates that these factors are highly predictive of heart disease in this dataset. Also note that chest pain (cp) is by far the most influential indicator for importance of detecting heart disease in this dataset. Also looking at the feature graph, sex, fasting blood sugar (fbs) and resting ecg have little importance in determining heart disease. Such insights can help focus further analysis on these critical factors, or even guide feature selection in future modeling.

Understanding feature importance enhances model interpretability, helping to explain why the model makes specific predictions. In applications like healthcare, this transparency is crucial for building trust in machine learning models.

Summary

This hands-on activity demonstrates how to train, evaluate, and interpret a decision tree for classifying heart disease data. Decision trees offer transparency, making it possible to understand which features the model considers most influential. As we close our discussion on decision trees and random forests, we've seen how these methods build upon simple decision-making processes to create powerful predictive models. Now, let's shift our focus from these ensemble methods to a different type of modeling that underpins much of modern AI: neural networks. In Chapter 11, we will delve into the fundamentals of neural networks and deep learning, exploring how these sophisticated architectures are designed to mimic the way the human brain processes information.

Below are reflective questions on this Chapter, Decision Trees and Random Forests:

Reflecting Questions:

1. **Understanding Basics**:

   - How does a decision tree use splitting to make predictions?

2. **Application of Concepts**:

   - Why is interpretability an important feature of decision trees, especially in fields like healthcare or finance?

3. **Model Strengths and Weaknesses**:

   - What are the major advantages of decision trees, and why might they not always perform well on new data?

4. **Key Techniques**:

- How do techniques like limiting tree depth or using a minimum number of samples per split prevent overfitting?

5. **Random Forest Benefits**:

   - How does randomness in feature selection and bootstrapping make Random Forests more robust than single decision trees?

6. **Feature Importance**:

   - Why is identifying feature importance useful when analyzing the results of a decision tree or Random Forest?

7. **Real-world Impact**:

   - Can you think of an example outside of healthcare where a decision tree might provide valuable insights?

8. **Ethical Considerations**:

   - How might bias in the dataset affect the decisions made by a model, and what can be done to reduce this bias?

9. **Algorithm Connections**:

   - How do you think Random Forests address the "instability" problem inherent in decision trees?

10. **Critical Thinking**:

   - What are some trade-offs you might encounter when choosing between a decision tree and a Random Forest for a specific problem?

Python Challenge 1: Train and Evaluate a

Decision Tree

Challenge:

Using the *Heart Disease* dataset, train a decision tree with a max_-depth of 4. Split the dataset into training and testing sets, then calculate and print the model's accuracy and classification report.

```
1   import pandas as pd
2   from sklearn.model_selection import train_test_split
3   from sklearn.tree import DecisionTreeClassifier
4   from sklearn.metrics import accuracy_score, classificatio\
5   n_report
6
7   # Load the dataset
8   data = pd.read_csv("heart_disease.csv")
9
10  # Define features (X) and target variable (y)
11  # Assuming 'target' is the target column
12  X = data.drop(columns=['target'])
13  y = data['target']
14
15  # Split the dataset into training and testing sets
16  X_train, X_test, y_train, y_test = train_test_split(X, y,\
17   test_size=0.2, random_state=42)
18
19  # Train the decision tree classifier
20  # <create and fit the model here>
21
22  # Make predictions and evaluate the model
23  # y_pred = <generate prediction here>
24  accuracy = accuracy_score(y_test, y_pred)
25  print("Accuracy:", accuracy)
```

```
26  print("\nClassification Report:\n", classification_report\
27  (y_test, y_pred))
```

Python Challenge 2: Visualize a Decision Tree

Challenge:
Train a decision tree and visualize it using plot_tree from sklearn.
Use the *Heart Disease* dataset and set max_depth=3.

```
1  import matplotlib.pyplot as plt
2  from sklearn.tree import plot_tree
3
4  # Train the decision tree model
5  # <Train and fit the decision tree here>
6
7  # Visualize the decision tree
8  # <Plot the tree here>
```

Python Challenge 3: Calculate and Plot Feature Importance

Challenge:
Train a decision tree on the *Heart Disease* dataset and create a bar chart of the feature importances to see which factors are most significant.

```
 1  # Train the decision tree model
 2  # <train and fit the decision tree model>
 3
 4  # Get feature importances
 5  feature_importances = # < Get feature importance
 6                        #   from the tree_model>
 7  feature_names = # <extract feature names from
 8                  # the training data>
 9
10
11  # Create a DataFrame for visualization
12  importance_df = pd.DataFrame({'Feature':
13      feature_names, 'Importance':
14      feature_importances})
15  importance_df = importance_df.sort_values
16      (by='Importance', ascending=False)
17
18  # Plot the feature importances
19  plt.figure(figsize=(10, 8))
20  plt.barh(importance_df['Feature'],
21      importance_df['Importance'], color='skyblue')
22  # Invert y-axis for descending order
23  plt.gca().invert_yaxis()
24
25  plt.xlabel('Importance')
26  plt.ylabel('Feature')
27  plt.title('Feature Importance in Predicting Heart
28      Disease')
29  plt.show()
```

Chapter 11: Neural Networks and Deep Learning

Introduction to Neural Networks and Basic Architecture

Neural networks represent a radical departure from traditional programming paradigms, introducing systems that learn from data. At its core, a neural network is a series of algorithms that endeavors to recognize underlying relationships in a set of data through a process that mimics the way the human brain operates. This chapter introduces you to the foundational concepts of neural networks and their basic architecture, which includes input layers, hidden layers, and output layers.

Input Layer: The input layer is the initial phase of the processing cycle. Here, individual neurons represent raw data. For instance, in image processing, each neuron in the input layer might correspond to the pixel intensity in different parts of an image.

Hidden Layers: Between the input layer and the output layer are one or more layers of neurons called hidden layers. These layers are termed "hidden" because they are not directly exposed to the input or output. Each neuron in these layers receives inputs from the neurons of the previous layer, processes the inputs, and passes the output to the next layer. The complexity and capability of a neural network increase with the number of hidden layers it contains, a concept known as depth.

Output Layer: The final layer, known as the output layer, produces the output of the neural network. For a classification task, the output layer would typically have as many neurons as there are classes, with each neuron representing the probability that a given input belongs to a specific class.

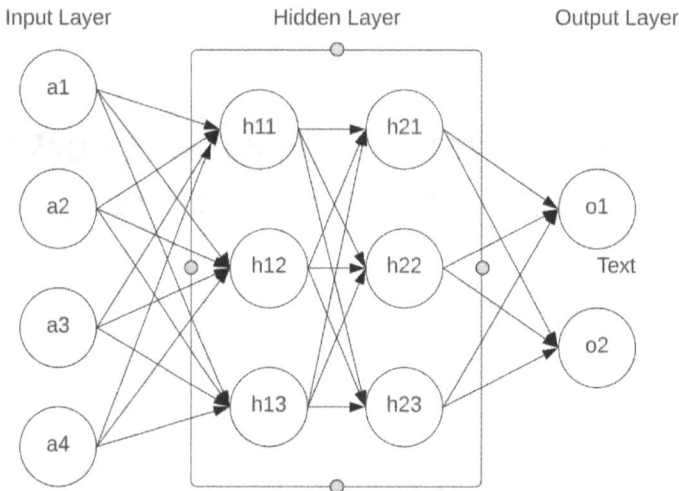

Figure 54. A Simple Neural Network

Deep Learning

What is Deep Learning? How is it Different from Traditional Machine Learning?

Deep learning is a subset of machine learning that has gained significant attention due to its ability to process large volumes of data and automatically extract complex patterns that are challenging for humans or traditional machine learning techniques. This section explores what deep learning is and how it differs from traditional machine learning.

Understanding Deep Learning

Deep learning models, particularly neural networks, are inspired by the structure and function of the brain called artificial neural networks. These models are composed of layers of interconnected nodes or "neurons," each layer designed to recognize different features of the data. The model's depth, characterized by many layers, allows it to learn from a vast amount of data, adapting to a variety of tasks, from image and speech recognition to natural language understanding.

Deep learning automates much of the feature extraction portion of the process, eliminating the need for manual segmentation of the data. Instead, these models learn to identify features directly from the data, training on raw input data and applying layers of learning to iteratively improve their accuracy.

How Deep Learning Differs from Traditional Machine Learning

1. Capability to Handle Unstructured Data:

   - **Traditional Machine Learning**: Traditional algorithms struggle with unstructured data, such as images, audio, and text. They typically require careful engineering and domain expertise to design feature extractors that transform raw data into a suitable internal representation or feature vector.
   - **Deep Learning**: Deep learning algorithms, by contrast, excel at working with unstructured data. Using multiple layers, they automatically learn the features directly

from the data, without the need for manual feature extraction.

2. **Scalability and Performance**:

- **Traditional Machine Learning**: These models often plateau in performance after reaching a certain amount of training data or complexity. More data or more complex features don't necessarily improve the performance of traditional machine learning algorithms.
- **Deep Learning**: The performance of deep learning models often improves as the size of the data increases. The layered architecture enables these models to scale with data and complexity, often outperforming traditional models when it comes to large and complex datasets.

3. **Hardware Dependencies**:

- **Traditional Machine Learning**: Traditional algorithms can often run on low-end machines without the need for extensive computational resources.
- **Deep Learning**: Deep learning models, particularly those involving large neural networks, require significant computational power. They often depend on GPUs or specialized hardware for efficient training and inference, which can be a limiting factor in their deployment.

4. **Interpretability and Transparency**:

- **Traditional Machine Learning**: Algorithms like decision trees, linear/logistic regression, etc., are often easier to interpret and understand. The decisions made by these models can be traced back clearly and explained logically.

- **Deep Learning**: One of the biggest challenges with deep learning models is their "black box" nature. The decisions made by deep neural networks can be difficult to interpret, as the learning process involves complex transformations across many layers.

Conclusion

While deep learning offers considerable advantages in terms of power and flexibility, it's not without its challenges. The choice between deep learning and traditional machine learning ultimately depends on the specific needs of the application, the availability of computational resources, and the required transparency of the model. Deep learning is best suited for applications involving large amounts of data and complex problems where traditional machine learning falls short. However, for simpler tasks, or when interpretability is a priority, traditional machine learning might be more appropriate.

Overview of Frameworks like TensorFlow and PyTorch

In the world of deep learning, the frameworks that stand out the most are TensorFlow and PyTorch. These tools have revolutionized how developers build, train, and deploy neural networks and other machine learning models. This section provides an overview of these two popular frameworks, discussing their core features, similarities, differences, and typical use cases.

TensorFlow

TensorFlow is an open-source library developed by Google Brain in 2015. It is designed to facilitate building and training machine learning models with a focus on deep learning. TensorFlow is known for its flexibility, scalability, and comprehensive ecosystem of tools and libraries that support machine learning development.

- **Key Features:**

  - **Graph Execution:** TensorFlow models consist of a computational graph where nodes represent operations, and edges represent the tensors (data) communicated between them. This allows TensorFlow to optimize computation, making use of parallel processing and making it highly efficient on large-scale systems.
  - **Eager Execution:** Introduced in later versions, eager execution allows operations to be evaluated immediately without building graphs. This makes the framework more interactive and intuitive, especially for beginners.
  - **TensorBoard:** TensorFlow provides TensorBoard, a tool for visualization that makes it easier to understand, debug, and optimize the models.
  - **Deployment and Scalability:** TensorFlow supports deployment on a variety of platforms, including servers, edge devices, and mobile applications, making it versatile for production environments.

- **Typical Use Cases:**

  - Large-scale machine learning deployments
  - Industrial applications where robustness and scalability are critical
  - Research and development requiring complex computational graph manipulations

PyTorch

PyTorch, developed by Facebook's AI Research lab, has grown in popularity due to its ease of use, dynamic computation graph, and strong community support. Released in 2016, PyTorch is particularly favored for academic and research-oriented projects thanks to its simplicity and flexibility.

- **Key Features:**

    - **Dynamic Computation Graph**: Known as autograd, PyTorch's dynamic nature allows changes to be made on-the-fly using imperative programming. It is particularly useful for models where conditions and loops make the network architecture dynamic.
    - **Pythonic Nature**: PyTorch seamlessly integrates with the Python data science stack, making it not only easy to learn but also robust and clear to debug.
    - **TorchScript**: PyTorch provides TorchScript, a way to create serializable and optimizable models from PyTorch code. This facilitates the easy transition from research to production.
    - **Extensive Libraries**: Similar to TensorFlow, PyTorch offers libraries like TorchVision for computer vision tasks, making it easy to implement and extend complex models.

- **Typical Use Cases:**

    - Rapid prototyping in research
    - Applications requiring complex, variable computational graphs
    - Projects where development speed and flexibility are more critical than immediate deployment at scale

FastAI

FastAI is an open-source library designed to democratize deep learning by making it more accessible and easier to use. Developed by Jeremy Howard and Rachel Thomas through the Fast.AI organization, it simplifies the process of obtaining state-of-the-art results without requiring the in-depth mathematical understanding typically associated with building deep learning models.

- Key Features:

  - **High-Level Abstractions**: FastAI offers higher-level components that can automatically handle many of the detailed settings and training procedures that can be barriers for new practitioners.
  - **Built on PyTorch**: Leveraging PyTorch's flexibility and power, FastAI extends it with more pre-built structures to speed up the coding process.
  - **Education-Focused**: With strong ties to the FastAI courses, the library is particularly well-suited for learners and educators, making it easier to teach and learn deep learning.
  - **Community and Ecosystem**: FastAI promotes an inclusive and practical approach to deep learning, supported by a vibrant community that contributes to continuous improvements and extensive documentation.

- Typical Use Cases:

  - Ideal for educators and students in the field of AI who are looking to grasp the practical aspects of deep learning quickly.
  - Useful for researchers and developers who need to prototype state-of-the-art models rapidly.
  - Great for projects where ease of use and speed of development are prioritized over custom architecture design.

Integration in the Learning Curriculum

Including FastAI in the curriculum can help bridge the gap between theoretical understanding and practical application. It allows students to quickly see results and thus stay motivated through:

- **Hands-on Practice:** FastAI's design lets students build and train models with fewer lines of code, which is encouraging for beginners.
- **Focus on Practical Results:** FastAI incorporates many best practices by default, helping students achieve impressive results early in their learning process.

When discussing TensorFlow, PyTorch, and now FastAI, it's beneficial to frame them not just as alternatives but as tools suited for different aspects of the learning and development journey in AI and machine learning:

- **TensorFlow** is ideal for large-scale deployments and production environments.
- **PyTorch** offers flexibility and power for research and development.
- **FastAI** excels in educational settings and rapid prototyping, with less upfront learning curve and more immediate results.

Having explored the different frameworks available for deep learning such as TensorFlow, PyTorch, and FastAI, it's now time to put theory into practice. In the next section, we will dive into a hands-on activity using one of the most well-known datasets in the machine learning community—the MNIST dataset. This dataset will provide a practical framework for understanding how deep learning models can be trained to recognize and interpret visual data.

Hands-On Activity: Using the MNIST Dataset to Identify Numbers

Introduction to MNIST Dataset

The MNIST dataset, which stands for Modified National Institute of Standards and Technology database, is a large collection of handwritten digits that is commonly used for training various image processing systems. The dataset contains 70,000 images of handwritten digits (0 through 9), each of which is 28x28 pixels. Each image is labeled with the digit it represents. This dataset is widely used for training and testing in the field of machine learning because it requires minimal data preprocessing and formatting, making it ideal for demonstrating the capabilities of neural network models to beginners.

Objective of the Next Section

In the upcoming section, we will develop a neural network model that uses the MNIST dataset to accurately recognize and classify handwritten digits. This practical application will serve as a concrete example to illustrate the process of:

Preparing data for a neural network. Building a simple neural network architecture suitable for image recognition. Training the model using one of the previously discussed frameworks. Evaluating the model's performance and tuning it for better accuracy. Learning Outcomes By the end of this hands-on activity, students will:

Understand the steps involved in preparing data for neural networks, including normalization and encoding. Gain practical experience in designing a neural network for a specific application— digit recognition. Learn how to train and fine-tune a neural network using real-world data. Develop skills in using a machine learning framework to implement a complete project from start to finish.

To create a practical example using PyTorch for classifying hand-written digits with the MNIST dataset, we'll go through setting up the environment, loading the data, building a simple neural network, training the model, and then evaluating its performance. This example will be structured to be easily followed in a classroom or self-learning setting.

Certainly! Switching to PyTorch for the MNIST example is a great choice, especially given its widespread use in both academia and industry. Below is a detailed section for your chapter that walks through building a simple neural network using PyTorch to classify images from the MNIST dataset. This guide will include everything from setting up your environment to training and evaluating the model.

MNIST Classification Using PyTorch

Setting Up Your Environment

First, ensure that PyTorch is installed in your environment. If not, you can install it by following the instructions on the official PyTorch website[1]. You'll also need torchvision, which helps with loading the MNIST dataset and transforming it.

```
1  pip install torch torchvision
```

Import Necessary Libraries

Start your Python script or Jupyter notebook by importing the required libraries:

---

[1]https://pytorch.org/get-started/locally/

```
1   import torch
2   from torch import nn, optim
3   from torchvision import datasets, transforms,
4       utils
5   from torch.utils.data import DataLoader
6   import matplotlib.pyplot as plt
7   import numpy as np
```

Load and Prepare the MNIST Dataset

The MNIST dataset contains images of handwritten digits (0-9) that are 28x28 pixels. We'll use PyTorch's torchvision to load the dataset.

```
1   # Define a transformation to normalize the data
2   transform = transforms.Compose
3       ([transforms.ToTensor(),
4
5       transforms.Normalize((0.5,), (0.5,))])
6
7   # Download and load the training data
8   trainset = datasets.MNIST(
9                   '~/.pytorch/MNIST_data/',
10      download=True,
11      train=True,
12      transform=transform)
13  trainloader = DataLoader(trainset, batch_size=64,
14      shuffle=True)
15
16  # Download and load the test data
17  testset = datasets.MNIST('~/.pytorch/MNIST_data/'
18      , download=True, train=False,
19      transform=transform)
20  testloader = DataLoader(testset, batch_size=64,
21      shuffle=False)
```

Displaying Some of the Dataset

Before diving into the complexities of training our neural network, it's crucial to familiarize ourselves with the actual data we will be using. Visualizing the dataset not only verifies that our data has been loaded and processed correctly but also provides us with an intuitive understanding of the challenges and characteristics of the task at hand. In this section, we will explore a subset of the MNIST dataset, which comprises thousands of handwritten digits. Each image is a grayscale representation of a digit, labeled with its corresponding numerical value. By examining these images, we can appreciate the variations and idiosyncrasies in handwriting styles that our model will need to learn. Let's begin by displaying a selection of these digit images in a grid format. This visual examination will help us assess the diversity and complexity of the dataset, setting a foundational understanding for the subsequent steps in our machine learning project."

```
1    # let's show some of the images in the dataset
2    import torchvision
3
4    def imshow(img):
5        # img is a torch tensor, so convert it to a
6    # numpy array after denormalization
7        img = img / 2 + 0.5      # unnormalize
8        npimg = img.numpy()
9        # Convert from Tensor image
10       plt.imshow(np.transpose(npimg, (1, 2, 0)))
11       # show the image
12       plt.show()
13
14   # Get some random training images
15   dataiter = iter(trainloader)
16   images, labels = next(dataiter)
17
18   # Show images in a 2x2 grid
```

```
19
20   # Set the figure size for better visibility
21   plt.figure(figsize=(10, 10))
22   imshow(torchvision.utils.make_grid(images[:4],
23       nrow=2))
24   plt.axis('off')   # Turn off axis numbers and ticks
25   plt.show()
```

Output

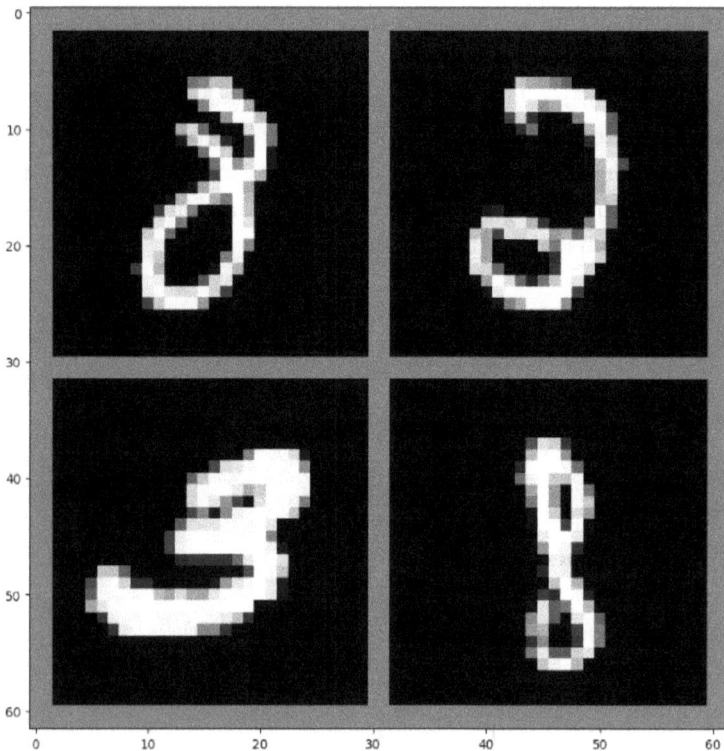

Figure 55. Sample MNIST Data

Note that the image set is not merely a collection of neatly printed
numbers; indeed, while many of the examples in the MNIST

dataset are easily interpreted by the human eye, some feature handwriting that could be ambiguous or challenging to decipher without context.

Building the Neural Network

We'll create a simple neural network with one hidden layer. PyTorch allows you to define models either using a class or sequentially. Here, we'll use a class for clarity.

Certainly! Introducing and explaining these neural network components clearly within your chapter is essential for readers, especially if they're new to machine learning or neural network architectures. Here's how you can introduce and explain these elements in a way that integrates seamlessly into your chapter on neural networks using the PyTorch framework:

Introducing Key Neural Network Components

To help our readers understand the architecture of the neural network we're using to classify handwritten digits in the MNIST dataset, let's discuss some of the fundamental components: linear layers, the Sigmoid function, and the Softmax function. Each plays a critical role in processing the input data and producing a usable output.

Linear Layers

The nn.Linear module in PyTorch creates a linear transformation to incoming data, ($y = xA^T + b$):

- **self.output = nn.Linear(256, 10)**: This line defines a linear layer that takes an input with 256 features and outputs 10 features. These 10 output features correspond to our 10 classes (digits 0 through 9). This layer is typically used at the output stage of a classification network, where each feature represents a class's "score" before final classification.

Activation Functions

Activation functions are crucial in neural networks as they introduce non-linearities into the model, allowing it to learn more complex patterns in the data.

- **Sigmoid Function**

    - `self.sigmoid = nn.Sigmoid()`: The Sigmoid function maps the input (any real-valued number) to an output value between 0 and 1. It's a form of the logistic function and is often used for models where we need to predict the probability as an output since the probability of anything exists only between the range of 0 and 1.

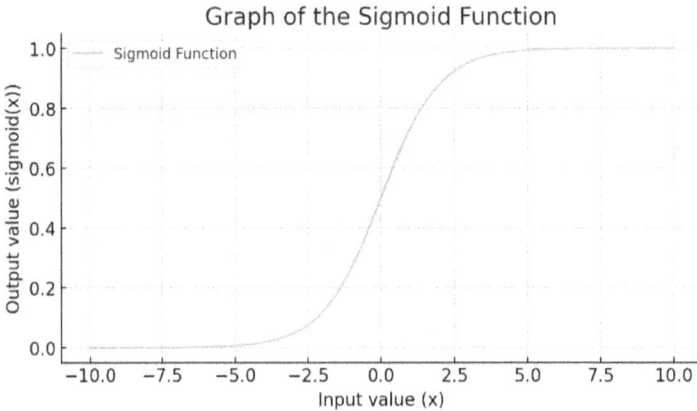

Figure 56. Sigmoid Function

Fun Fact: The Sigmoid Function and Human Nerve Cells

The sigmoid function, often used in neural networks as an activation function, has a biological counterpart in the way human nerve cells (neurons) operate. In neuroscience, the sigmoid function mimics the activation pattern of neurons: it represents how neurons fire. Just as the sigmoid function gradually transitions from 0 to 1, a neuron fires gradually, increasing the intensity of its signal once the input exceeds a certain threshold. This biological analogy isn't just a coincidence; it inspired the early development of neural networks, aiming to mimic how the human brain processes information.

- **Softmax Function**

  - `self.softmax = nn.LogSoftmax(dim=1)`: While the Sigmoid function is suitable for binary classification, the Softmax function is used for multi-class classification tasks. It converts a vector of values to a probability distribution, where the probabilities of each value are proportional to the exponentials of the input numbers. The `LogSoftmax` is a variation that applies the logarithm after the Softmax transformation, which often helps in improving numerical stability and performance during the training phase.

Example Usage in a Neural Network

These components are typically used as follows in a network:

- After passing input through multiple layers (e.g., linear transformations and non-linear activations), the final predictions are formed using a linear layer followed by a Softmax function if the task is a multi-class classification.

- The Sigmoid or Softmax output can then be used to calculate loss during training and make predictions during inference.

```
class MNISTNetwork(nn.Module):
    def __init__(self):
        super(MNISTNetwork, self).__init__()
        # 28x28 = 784 input pixels, 256 outputs
        self.hidden = nn.Linear(784, 256)
        # 10 output units for 10 classes (0-9)
        self.output = nn.Linear(256, 10)
        self.sigmoid = nn.Sigmoid()
        self.softmax = nn.LogSoftmax(dim=1)

    def forward(self, x):
        x = x.view(x.shape[0], -1)  # Flatten the
    images
        x = self.hidden(x)
        x = self.sigmoid(x)
        x = self.output(x)
        x = self.softmax(x)
        return x

# Create the network
model = MNISTNetwork()
```

Defining the Loss Function and Optimizer

To train a neural network effectively, we need two crucial components: a loss function and an optimizer. Let's delve deeper into these components, focusing on the use of cross-entropy loss for multi-class classification tasks like MNIST and stochastic gradient descent (SGD) as an optimization strategy.

Cross-Entropy Loss

Cross-Entropy Loss is a performance metric that quantifies the difference between two probability distributions - the predicted probability distribution output by the model and the actual distribution of the labels. In the context of classification, it measures the discrepancy between the predicted class probabilities and the actual class labels. This loss function is particularly suited for classification problems including those involving multiple classes, where it helps to drive the training process by penalizing predictions that diverge from the actual labels.

- **Mathematical Formulation**: If you have a model that outputs a probability (p) for each class and a true label (y) that is 1 for the correct class and 0 otherwise, the cross-entropy loss for one sample can be calculated as:

  [L = -\sum_{c=1}^C y_c \log(p_c)]

  where (C) is the number of classes, (y_c) is a binary indicator (0 or 1) if class label (c) is the correct classification for the observation, and (p_c) is the predicted probability that the observation belongs to class (c).

Stochastic Gradient Descent (SGD)

Imagine you're standing at the top of a hilly mountain covered in fog, and your goal is to reach the lowest point of the mountain— the valley below. However, because of the fog, you can't see the entire path leading directly to the valley. So, you must feel your way down step by step.

Feeling the Slope: Each step you take, you can feel whether the ground is sloping up or down under your feet. This is similar to calculating the "gradient" in gradient descent. The gradient tells you the slope of the mountain under your current position.

Deciding the Direction: Based on the slope, you decide which way to step to go downwards. In gradient descent, this is like adjusting your parameters (or model's weights) slightly in the direction that reduces your error, or the height from the valley.

Taking a Step: You then take a careful step in that direction. The size of your step is important—if you step too far, you might overshoot and miss the valley; if you step too little, it might take you a long time to get there. In gradient descent, this step size is controlled by a parameter called the "learning rate."

Reassess and Repeat: After each step, you stop to feel the slope again, deciding the next direction to step. You repeat this process, gradually making your way downhill, adjusting your path as you gather more information about the slope under your feet with each step.

Approaching the Goal: As you get closer to the valley, your steps become smaller and more cautious because the slopes become less steep. This is akin to the gradient getting smaller as the parameters start to converge to the optimal values that give you the lowest point on the mountain—your target.

In summary, gradient descent is a method where you make iterative adjustments (steps) to your parameters (directions), aiming to minimize errors (reach the valley) in your model (journey). Each step is determined by the slope (gradient) at your current position, guiding you towards your goal even though you can't see the entire path from the start. This process helps ensure that even in a complex, multidimensional space, you can find the best solution to your problem, just like finding the lowest point on a foggy mountain.

Stochastic Gradient Descent is an optimization method used to minimize the loss function by iteratively adjusting the parameters of the model. SGD updates the parameters using a randomly selected subset of data rather than the full dataset, which makes the updates faster and reduces the computational burden.

- **Optimization Process**: The basic idea behind SGD involves taking the gradient (or an estimate of the gradient) of the loss function with respect to the model parameters, and then updating the parameters in the opposite direction of the gradient. For each training example (x_i) with label (y_i), the parameters (\theta) are updated as follows:

  [\theta = \theta - \eta \cdot \nabla_\theta J(\theta; x_i, y_i)]

  where (\eta) is the learning rate, and (\nabla_\theta J) is the gradient of the loss function (J) with respect to the parameters (\theta).

- **Learning Rate**: The learning rate (\eta) is a hyperparameter that controls how much the parameters change on each update. This value needs to be set carefully, as too high a value can cause the model to converge too quickly to a suboptimal solution, and too low a value can slow down the convergence process.

Integration in Training

When training a neural network, we typically define the loss function and optimizer in the setup phase and then use them during the training loop. Here's how you might see these components configured in a PyTorch setup:

By using cross-entropy loss and stochastic gradient descent, we aim to train a model that not only performs well on the training data but also generalizes to new, unseen data effectively.

```
1   # loss function
2   criterion = nn.NLLLoss()
3   # optimizer that uses stochastic gradient decent
4   optimizer = optim.SGD(model.parameters(), lr=0.03)
```

Train the Network

Now, train the network using the training data from the MNIST set, output the loss after each epoch:

```
epochs = 15
for e in range(epochs):
    running_loss = 0
    for images, labels in trainloader:
        optimizer.zero_grad()

        output = model(images)
        loss = criterion(output, labels)

        loss.backward()
        optimizer.step()

        running_loss += loss.item()
    else:
        print(f"Training loss: {running_loss/len
    (trainloader)}")
```

Output Training loss: 1.682666209651463 Training loss: 0.8168541670862292 Training loss: 0.5716605792358231 Training loss: 0.4758968580760427 Training loss: 0.42519174487606043 Training loss: 0.39362197577444985 Training loss: 0.37197302788623104 Training loss: 0.35584060105878407 Training loss: 0.34312510461822504 Training loss: 0.3330684964463655 Training loss: 0.32436909250168405 Training loss: 0.31692451578594727 Training loss: 0.3101521505634668 Training loss: 0.3042210640906017 Training loss: 0.29875197523692526

The training loss values provided illustrate the neural network's learning process over successive epochs. Initially, the loss started relatively high at 1.6827, indicating that the model's predictions were quite far from the actual target values. As training progressed, the loss consistently decreased, reflecting improvements in the model's accuracy and its ability to generalize from the training data.

By the second epoch, the loss had significantly reduced to 0.8169, showing a substantial improvement as the model adjusted its weights and biases to better capture the underlying patterns in the data. This trend of decreasing loss continued, albeit at a diminishing rate, indicating that each additional adjustment to the model's parameters provided a smaller incremental benefit.

By the mid-training point, the loss values had settled into a more gradual decline, dropping to around 0.4252 by the fifth epoch and further to about 0.2988 by the fifteenth epoch. This slowing rate of decrease suggests that the model was approaching its optimal state given the current architecture and training data.

The continual but smaller reductions in loss from epoch ten through fifteen, from 0.3331 to 0.2988, imply that the model was fine-tuning its parameters, squeezing out marginal gains in performance. This phase is critical as it often leads to a more robust model that performs well on unseen data, assuming no overfitting occurs.

Overall, the descending pattern of training loss over these epochs is indicative of effective learning, where the model progressively learns and adapts, reducing the gap between the predicted outputs and the actual values with each epoch.

Evaluate the Model

Let's test the accuracy of your model on the test dataset:

```
1    # Initialize counters for correct predictions and
2    # total predictions
3    correct_count, all_count = 0, 0
4
5    # Iterate through each batch of images and labels
6    # in the test dataset
7    for images, labels in testloader:
8        # Iterate through each image and
9    # corresponding label in the batch
10       for i in range(len(labels)):
11           # Disable gradient calculation to speed
12   # up the process and reduce memory usage
13           with torch.no_grad():
14   # Model inference: pass the image
15   # through the model to get the predicted
16   # probabilities for each class
17               # .view(1, 784) reshapes the image to
18   # the appropriate batch size and input dimensions
19   # required by the model
20               logps = model(images[i].view(1, 784))
21
22   # Convert the log probabilities to actual
23   # probabilities for easier interpretation
24           ps = torch.exp(logps)
25   # Convert the tensor to a numpy array and
26   # get the first item (since batch size is 1,
27   # there is only one item)
28           probab = list(ps.numpy()[0])
29   # Find the predicted label by finding the
30   # index of the highest probability
31           pred_label = probab.index(max(probab))
32   # Get the true label for the current
33   # image from the batch
34           true_label = labels.numpy()[i]
35   # Check if the predicted label matches
```

```
36    # the true label
37            if true_label == pred_label:
38    # If correct, increment the correct
39    # prediction count
40                correct_count += 1
41    # Always increment the total count
42            all_count += 1
43
44    # Print the total number of images tested
45    print("Number Of Images Tested =", all_count)
46    # Calculate and print the model accuracy
47    print("\nModel Accuracy =", (correct_count /
48        all_count))
```

Output Number Of Images Tested = 10000

Model Accuracy = 0.9196

Key Points:

1. **Disabling Gradient Calculation**: This is crucial during testing and evaluation since gradients are not needed for updating model parameters and can consume additional memory and processing.

2. **Reshaping the Image**: The .view(1, 784) operation reshapes the image test data to fit the input requirements of the model. It sets the batch size to 1 and flattens the 28x28 image into a 784-dimensional vector.

3. **Probability Conversion**: Converting log probabilities of the output to actual probabilities (torch.exp(logps)) simplifies finding the class with the highest probability.

4. **Accuracy Calculation**: By comparing predicted labels with true labels and counting matches, we calculate the overall accuracy of the model over the test set.

Conclusion

This section has walked you through setting up a basic neural network in PyTorch to classify handwritten digits from the MNIST dataset. Through this exercise, students can learn how neural networks are structured, trained, and used for real-world applications.

Here are a few reflective questions to help review your understanding of neural network technologies in Python:

Reflective Questions for Chapter 11

1. **Conceptual Understanding**:

   - What is the significance of the input, hidden, and output layers in a neural network?
   - How does deep learning differ from traditional machine learning in terms of handling unstructured data?
   - Why are activation functions necessary in a neural network?

2. **Framework Selection**:

   - What are the advantages of using PyTorch for developing neural networks?
   - When would you prefer TensorFlow over PyTorch, and vice versa?

3. **Practical Considerations**:

   - How does increasing the depth of a neural network impact its performance and computational requirements?
   - Why might you choose to use stochastic gradient descent (SGD) over other optimization methods?

Python Challenges for Chapter 11

Here's how we can adapt the challenges for the **Penguins Dataset** using PyTorch.

Challenge 1: Build a Neural Network

Task: Build a simple neural network in PyTorch to classify penguins into one of three species: Adelie, Chinstrap, or Gentoo.

```
import torch
from torch import nn

class PenguinNetwork(nn.Module):
    def __init__(self):
        super(PenguinNetwork, self).__init__()
        # create a input/hidden linear layer
        # with 4 inputs
        # Input layer (4 features),
        # Hidden layer (16 neurons)
        self.hidden = # <create layer here>
        # linear output layer, 16 inputs,
        #  3 outputs
        # (3 penguin classes in the output)
        self.output = # <create output layer
                      # here>

        # Activation for hidden layer
        self.relu = nn.ReLU()

        # Softmax for multiclass classification
        self.softmax = nn.Softmax(dim=1)
```

```
23
24
25     def forward(self, x):
26         x = self.hidden(x)
27         x = self.relu(x)
28         x = self.output(x)
29         x = self.softmax(x)
30         return x
31
32     # Instantiate the model
33     model = PenguinNetwork()
34     print(model)
```

Challenge 2: Train the Neural Network on Penguins Dataset

Task: Preprocess the Penguins dataset, train the neural network, and observe the loss.

```
1    from sklearn.model_selection import
2        train_test_split
3    from sklearn.preprocessing import StandardScaler,
4        LabelEncoder
5    import pandas as pd
6    import seaborn as sns
7    from torch.utils.data import DataLoader,
8        TensorDataset
9    import torch.optim as optim
10
11   # Load the dataset
12   # Drop rows with missing values
```

```
13    penguins = sns.load_dataset("penguins").dropna()
14
15    # Select features and target
16    X = penguins[["bill_length_mm", "bill_depth_mm",
17        "flipper_length_mm", "body_mass_g"]].values
18    y = penguins["species"].values
19
20    # Encode target labels
21    label_encoder = LabelEncoder()
22    y = label_encoder.fit_transform(y)  # Converts
23        species names to integers (0, 1, 2)
24
25    # Split into training and test datasets
26    X_train, X_test, y_train, y_test =
27        train_test_split(X, y, test_size=0.3,
28        random_state=42)
29
30    # Normalize features
31    scaler = StandardScaler()
32    X_train = scaler.fit_transform(X_train)
33    X_test = scaler.transform(X_test)
34
35    # Convert to PyTorch tensors
36    X_train = torch.tensor(X_train,
37        dtype=torch.float32)
38    y_train = torch.tensor(y_train, dtype=torch.long)
39        # Long type for classification
40    X_test = torch.tensor(X_test, dtype=torch.float32)
41    y_test = torch.tensor(y_test, dtype=torch.long)
42
43    # Create DataLoaders
44    train_dataset = TensorDataset(X_train, y_train)
45    train_loader = DataLoader(train_dataset,
46        batch_size=16, shuffle=True)
47
```

```
48    # Define the model, loss function, and optimizer
49    # Create your model from the previous challenge.
50    model = # <create the model here>
51    criterion = nn.CrossEntropyLoss()
52    # Cross-Entropy for multiclass classification
53    optimizer = optim.Adam(model.parameters(),
54        lr=0.01)
55
56    # Train the model
57    epochs = 20
58    for epoch in range(epochs):
59        running_loss = 0.0
60        for data, labels in train_loader:
61            optimizer.zero_grad()
62            outputs = model(data)
63            loss = criterion(outputs, labels)
64            loss.backward()
65            optimizer.step()
66            running_loss += loss.item()
67        print(f"Epoch {epoch+1}, Loss: {running_loss
68        /len(train_loader):.4f}")
```

Challenge 3: Evaluate the Model

Task: Test the trained model on the test dataset and calculate its accuracy for predicting the penguin species.

Solution:

```
1   # Evaluate the model
2   correct = 0
3   total = 0
4
5   with torch.no_grad():
6       outputs = model(X_test)
7       # Get class with highest probability
8       _, predictions = torch.max(outputs, 1)
9
10      correct = (predictions == y_test).sum().item()
11      total = y_test.size(0)
12
13  accuracy = # <compute the overall accuracy here>
14  print(f"Validation Accuracy: {accuracy *
15      100:.2f}%")
```

Chapter 12: Evaluating Machine Learning Models

In the realm of machine learning, building a model is only part of the journey. Evaluating its performance accurately and tuning it for better results are critical steps that determine the effectiveness of a machine learning solution in real-world applications. This chapter focuses on the metrics used to evaluate classification models, explaining how each metric provides insights into the model's performance and how they can guide the tuning process to achieve more reliable outcomes.

Metrics for Classification: Accuracy, Precision, Recall, F1 Score

Accuracy

Accuracy is the simplest and most intuitive performance metric for classification models. It measures the proportion of correct predictions (both true positives and true negatives) among the total number of cases examined. To calculate accuracy, simply divide the number of correct predictions by the total number of predictions made:

[\text{Accuracy} = \frac{\text{Number of Correct Predictions}}{\text{Total Number of Predictions}}]

While accuracy is useful, it can be misleading in situations where class distributions are imbalanced. For instance, if 95% of the data points belong to one class, a model can achieve 95% accuracy by trivially predicting the majority class for all inputs.

Precision

Precision is the ratio of correctly predicted positive observations to the total predicted positives. It is a crucial metric when the costs of False Positives are high. For example, in spam detection, a false positive (labeling a legitimate email as spam) is more disruptive than a false negative (failing to detect a spam email).

$$\text{Precision} = \frac{\text{True Positives}}{\text{True Positives} + \text{False Positives}}$$

Recall (Sensitivity)

Recall, also known as sensitivity or the true positive rate, measures the ability of a model to find all the relevant cases (positive examples). It is the ratio of correctly predicted positive observations to all observations in the actual class. Recall is particularly important when the cost of False Negatives is high. For instance, in disease screening, failing to detect a sickness (a false negative) can be more critical than falsely detecting the disease (a false positive).

$$\text{Recall} = \frac{\text{True Positives}}{\text{True Positives} + \text{False Negatives}}$$

F1 Score

The F1 Score is the harmonic mean of Precision and Recall. It is a way to combine both precision and recall into a single measure that captures both properties. This score is particularly useful when you need to balance precision and recall and there is an uneven class distribution (large number of actual negatives).

[\text{F1 Score} = 2 \times \left(\frac{\text{Precision} \times \text{Recall}}{\text{Precision} + \text{Recall}}\right)]

Summary

This section has introduced four fundamental metrics for evaluating the performance of classification models. Each metric provides different insights into the strengths and weaknesses of a model. Accurately measuring performance using these metrics is crucial for diagnosing model behaviors and guiding the subsequent tuning of model parameters to address specific needs of the application. In the following sections, we will delve deeper into how these metrics can be used in practice to evaluate and refine machine learning models, ensuring they perform optimally when deployed in real-world scenarios.

Metrics Example

Let's consider a simple example where we have predictions from a model and the corresponding actual labels. We will calculate Accuracy, Precision, Recall, and the F1 Score using Python, particularly with the help of the `sklearn.metrics` library which provides straightforward functions to compute these metrics.

Setup Example Data

First, we'll set up an example with binary classification results. Assume the following predictions and true labels:

```
1   from sklearn.metrics import accuracy_score,
2       precision_score, recall_score, f1_score,
3       confusion_matrix
4
5   # Example binary classification outcomes
6
7   # Actual labels
8   y_true = [0, 1, 1, 0, 1, 0, 1, 0, 0, 1]
9    # Predicted labels by the model
10  y_pred = [0, 1, 0, 0, 1, 0, 1, 1, 0, 0]
```

Calculating Metrics

1. Accuracy

Accuracy measures the overall correctness of the model and is the simplest metric.

```
1   accuracy = accuracy_score(y_true, y_pred)
2   print(f"Accuracy: {accuracy:.2f}")
```

In this example 7 of the predicted and actual values match out of 10, so that's an accuracy of 7/10 or .70.

2. Precision

Precision is a metric that measures the accuracy of the positive predictions made by a model. Specifically, it quantifies the proportion of positive identifications that were actually correct. In simpler terms, precision answers the question: "Out of all the instances the model labeled as positive, how many were actually positive?" This metric is particularly crucial in scenarios where the cost of a false positive is high, such as in spam detection, fraud alert systems, or disease screenings, where false alarms can have serious consequences. Precision ensures that a model is not only accurate but also reliable in its positive classifications. Let's step through the calculation.

Definitions Needed for Precision Calculation

- **True Positives (TP)**: These are cases where the model correctly predicts the positive class. In other words, the model identifies an observation as positive, and it is indeed positive.
- **False Positives (FP)**: These occur when the model incorrectly predicts the positive class. That is, the model identifies an observation as positive, but it is actually negative.

Steps to Calculate Precision

1. **Identify True Positives (TP)**: Compare each element of y_true and y_pred and count the instances where both are 1.

   - In your data:

     – At index 1, y_true is 1 and y_pred is 1.
     – At index 4, y_true is 1 and y_pred is 1.
     – At index 6, y_true is 1 and y_pred is 1.

   - So, you have **3 True Positives**.

2. **Identify False Positives (FP)**: Check where y_pred is 1 but y_true is 0.

   - In your data:

     – At index 7, y_true is 0 and y_pred is 1.

   - So, you have **1 False Positive**.

3. **Calculate Precision**: Precision is calculated using the formula: [\text{Precision} = \frac{\text{TP}}{\text{TP} + \text{FP}}]

- Plugging in the numbers from above: [\text{Precision} = \frac{3}{3 + 1} = \frac{3}{4} = 0.75]

The precision of your model based on the given data is 0.75, or 75%. This means that when the model predicts an observation as positive, it is correct 75% of the time.

This calculation can be easily done using Python with the help of the `precision_score` function from the `sklearn.metrics` package, which automates these calculations. Here is how you can do it programmatically:

```
from sklearn.metrics import precision_score

# Actual labels
y_true = [0, 1, 1, 0, 1, 0, 1, 0, 0, 1]
# Predicted labels
y_pred = [0, 1, 0, 0, 1, 0, 1, 1, 0, 0]

precision = precision_score(y_true, y_pred)
print(f"Calculated Precision: {precision}")
```

Output Calculated Precision: .75

This result matches our manual calculation of .75.

3. Recall

Recall, also known as sensitivity or the true positive rate, is a crucial metric in situations where it is essential to capture as many positives as possible, such as in medical diagnostics or fraud detection. It measures the ratio of correctly predicted positive observations to all actual positives. Let's step through the calculation of recall given your dataset.

Definitions Needed for Recall Calculation

- **True Positives (TP)**: Cases where the model correctly predicts the positive class.
- **False Negatives (FN)**: Cases where the model incorrectly predicts the negative class when it is actually positive.

Steps to Calculate Recall

1. **Identify True Positives (TP)**: Count the instances where both the actual label and the predicted label are 1.

   - In your data:

     - At index 1, y_true is 1 and y_pred is 1.
     - At index 4, y_true is 1 and y_pred is 1.
     - At index 6, y_true is 1 and y_pred is 1.

   - Thus, you have **3 True Positives**.

2. **Identify False Negatives (FN)**: Check where y_pred is 0 but y_true is 1.

   - In your data:

     - At index 2, y_true is 1 and y_pred is 0.
     - At index 9, y_true is 1 and y_pred is 0.

   - So, you have **2 False Negatives**.

3. **Calculate Recall**: Recall is calculated using the formula: $$\text{Recall} = \frac{\text{TP}}{\text{TP} + \text{FN}}$$

   - Plugging in the numbers from above: $$\text{Recall} = \frac{3}{3 + 2} = \frac{3}{5} = 0.60$$

Summary

The recall of your model based on the given data is 0.60, or 60%. This means that of all the actual positive cases, the model correctly identifies 60% of them.

This calculation can also be done using Python with the help of the recall_score function from the sklearn.metrics package, which automates these calculations. Here is how you can do it programmatically:

```
1  from sklearn.metrics import recall_score
2
3
4  # Actual labels
5  y_true = [0, 1, 1, 0, 1, 0, 1, 0, 0, 1]
6  # Predicted labels
7  y_pred = [0, 1, 0, 0, 1, 0, 1, 1, 0, 0]
8
9  recall = recall_score(y_true, y_pred)
10 print(f"Calculated Recall: {recall}")
```

Output Calculated Recall: .60

This code will output the recall which matches our manual calculation of .60.

4. F1 Score

The F1 Score is a crucial metric in classification, especially when you seek a balance between Precision and Recall. It is the harmonic mean of Precision and Recall, providing a single score that balances both the concerns of catching as many positives as possible (recall) and being right when it does predict a positive (precision).

Let's calculate the F1 Score manually using the same example data provided earlier and previously calculated precision and recall:

Calculate F1 Score: The F1 Score is the harmonic mean of Precision and Recall, calculated as: [\text{F1 Score} = 2 \times \left(\frac{\text{Precision} \times \text{Recall}}{\text{Precision} + \text{Recall}}\right) = 2 \times \left(\frac{0.75 \times 0.60}{0.75 + 0.60}\right)] [\text{F1 Score} = 2 \times \left(\frac{0.45}{1.35}\right) = 2 \times 0.3333 = 0.67]

Summary

The calculated F1 Score of approximately 0.67 indicates a moderate balance between precision and recall. This metric is particularly useful when you need to take both false positives and false negatives into account, providing a more balanced view of the model's performance, especially in datasets where class distribution might be imbalanced.

This method of calculation shows the importance of both precision and recall in determining the F1 Score and underscores its usefulness in situations where both types of errors have significant implications. Here is the calculation for f1 performed in python:

```
1   f1 = f1_score(y_true, y_pred)
2   print(f"F1 Score: {f1:.2f}")
```

Output F1 Score: 0.67

Complete Python Example

Here is how you can put together all the metrics we have learned in a single Python script:

```
1   from sklearn.metrics import accuracy_score,
2       precision_score, recall_score, f1_score
3
4   # Example binary classification outcomes
5
6   # Actual labels
7   y_true = [0, 1, 1, 0, 1, 0, 1, 0, 0, 1]
8
9   # Predicted labels by the model
10  y_pred = [0, 1, 0, 0, 1, 0, 1, 1, 0, 0]
11
12
13  # Calculate metrics
14  accuracy = accuracy_score(y_true, y_pred)
15  precision = precision_score(y_true, y_pred)
16  recall = recall_score(y_true, y_pred)
17  f1 = f1_score(y_true, y_pred)
18
19  # Print results
20  print(f"Accuracy: {accuracy:.2f}")
21  print(f"Precision: {precision:.2f}")
22  print(f"Recall: {recall:.2f}")
23  print(f"F1 Score: {f1:.2f}")
```

Confusion Matrix

Including a confusion matrix in your evaluation can provide a visual and detailed insight into the performance of your classification model, especially regarding how it classifies each class correctly or incorrectly. Here's a step-by-step guide on how to calculate and visualize a confusion matrix using Python, numpy, and seaborn for the provided dataset.

Data and Setup

Let's first ensure you have all necessary Python libraries installed and imported:

```
1   import numpy as np
2   import seaborn as sns
3   import matplotlib.pyplot as plt
4   from sklearn.metrics import confusion_matrix
```

Prepare the Data

You provided the data as:

```
1   # Actual labels
2   y_true = [0, 1, 1, 0, 1, 0, 1, 0, 0, 1]
3   # Predicted labels
4   y_pred = [0, 1, 0, 0, 1, 0, 1, 1, 0, 0]
```

Calculate the Confusion Matrix

First, use sklearn.metrics to calculate the confusion matrix:

```
1   cm = confusion_matrix(y_true, y_pred)
2   print(cm)
```

Output [[4 1] [2 3]]

Structure of the Confusion Matrix

The confusion matrix is structured as follows for a binary classification problem:

[\begin{array}{cc} \text{TN} & \text{FP} \ \text{FN} & \text{TP} \ \end{array}]

Where:

- **TP (True Positives)**: The cases in which the model correctly predicts the positive class.
- **TN (True Negatives)**: The cases in which the model correctly predicts the negative class.
- **FP (False Positives)**: The cases in which the model incorrectly predicts the positive class.
- **FN (False Negatives)**: The cases in which the model incorrectly predicts the negative class.

Your Confusion Matrix Explained

Given the matrix [[4, 1], [2, 3]]:

- **True Negatives (TN)**: 4

  – The model predicted the negative class correctly 4 times.

- **False Positives (FP)**: 1

  – The model predicted the positive class incorrectly once.

- **False Negatives (FN)**: 2

  – The model failed to identify the positive class correctly 2 times; it predicted them as negative.

- **True Positives (TP)**: 3

  – The model predicted the positive class correctly 3 times.

Visual Representation

The confusion matrix can be visualized as follows, which might help in understanding the layout:

[\begin{array}{cc|c|c|} & & \text{Predicted} & \text{Predicted} \ & & \text{Negative} & \text{Positive} \ \hline \text{Actual} & \text{Negative} & 4 & 1 \ \text{Negative} & \text{Positive} & 2 & 3 \ \hline \end{array}]

Implications

From the matrix:

- The model has a reasonably good ability to identify both classes, with a slightly better performance on predicting the negative class correctly.
- The number of False Negatives suggests that there might be a slight issue with sensitivity, as the model misses 2 positive cases. This could be critical depending on the application (e.g., failing to detect a disease in medical diagnostics).
- The relatively low number of False Positives indicates that the model is quite precise; it doesn't often label a negative sample as positive.

Visualizing the Confusion Matrix

Next, use seaborn to create a heatmap for the confusion matrix:

```
1  plt.figure(figsize=(8, 6))
2  sns.heatmap(cm, annot=True, fmt="d", cmap="Blues"
3      , cbar=False)  # "d" for decimal format
4  plt.xlabel('Predicted Labels')
5  plt.ylabel('True Labels')
6  plt.title('Confusion Matrix')
7  plt.show()
```

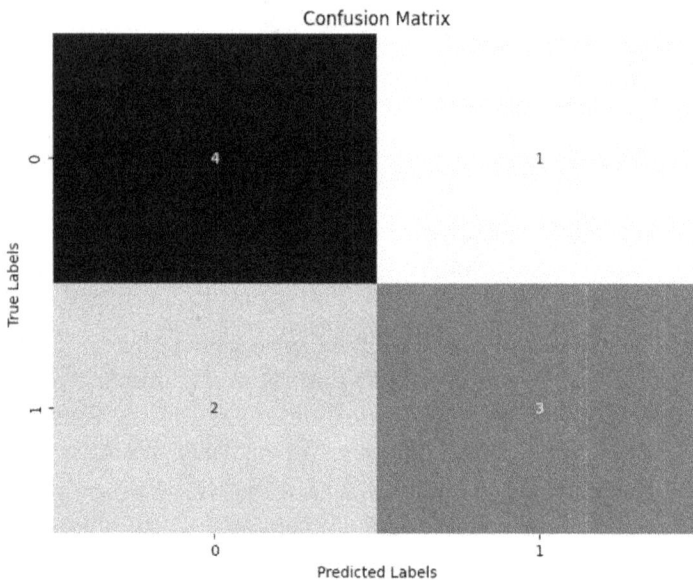

Figure 57. Confusion Matrix for Sample

This code block will:

- Calculate the confusion matrix between the true labels (y_-true) and the predictions (y_pred).
- Plot the matrix using seaborn's heatmap function.

    – annot=True makes sure the numbers are shown in each cell of the matrix.

– `fmt="d"` formats these numbers as integers (decimals without any decimal places).
– `cmap="Blues"` chooses a blue color scheme for the heatmap.

- Label the x-axis as "Predicted Labels" and the y-axis as "True Labels".
- Set the title of the heatmap as "Confusion Matrix".

Conclusion

The confusion matrix visually represents the accuracy of the classification model across all classes, showing the number of correct and incorrect predictions broken down by class. This provides intuitive insights into which classes are being predicted correctly or misclassified, facilitating a deeper understanding of model performance beyond scalar metrics like accuracy, precision, recall, and F1 score.

Metrics for Regression: Mean Squared Error and R-Squared

After exploring classification metrics, our focus shifts towards regression models, which predict continuous outcomes. To evaluate the performance of these models, we primarily use two metrics: Mean Squared Error (MSE) and R-squared (R^2). Both provide insights into how well a model captures the variance in the dataset and how close the predicted values are to the actual values.

Mean Squared Error (MSE)

Mean Squared Error is a widely used metric for quantifying the accuracy of a regression model. It measures the average of the squares of the errors—that is, the average squared difference between the estimated values and the actual value.

- **Formula**:

  [\text{MSE} = \frac{1}{n} \sum_{i=1}{^n} $(y\_i - \hat{y}\_i)2$]

  Where (y_i) is the actual value, (\hat{y}_i) is the predicted value from the model, and (n) is the number of observations.
- **Interpretation**:

  A lower MSE indicates a better fit of the model to the data, as it suggests smaller differences between the predicted and actual values. MSE is always non-negative, and values closer to zero are ideal.

R-Squared (Coefficient of Determination)

R-Squared measures the proportion of the variance in the dependent variable that is predictable from the independent variables. It provides an indication of goodness of fit and thus a measure of how well unseen samples are likely to be predicted by the model.

- **Formula**:

  [$R^2 = 1$ - \frac{\text{Sum of Squares of Residuals}}{\text{Total Sum of Squares}}

  ] [R2 = 1 - \frac{\sum_{i=1}{^n} $(y\_i - \hat{y}\_i)2$}{\sum_{i=1}{^n} $(y\_i$ - \overline{y})2}]

  Where (\overline{y}) is the mean of the observed data (y_i), (\hat{y}_i) are the predicted values by the model, and (y_i) are the actual values.
- **Interpretation**:

  R^2 is a statistical measure that represents the proportion of the variance for a dependent variable that's explained by an independent variable or variables in a regression model. An R^2 of 1 indicates that the regression predictions perfectly fit the data. Values of R^2 closer to 1 are generally considered better, indicating that more of the variance is accounted for by the model.

Practical Example in Python

Let's illustrate how to calculate these metrics in Python using a simple linear regression model:

```
from sklearn.metrics import mean_squared_error,
    r2_score
from sklearn.linear_model import LinearRegression
from sklearn.model_selection import
    train_test_split
import numpy as np

# Generating larger example data
# Array of 100 values with mean = 1.5,
# stddev = 2.5
np.random.seed(0)
X = 2.5 * np.random.randn(100) + 1.5

# Generate 100 residual terms
res = 0.5 * np.random.randn(100)

# Actual values of Y
y = 2 + 0.3 * X + res


# Reshape X to be a 2D array
X = X.reshape(-1, 1)
# Split the data
X_train, X_test, y_train, y_test =
    train_test_split(X, y, test_size=0.2,
    random_state=0)
```

```
27
28   # Fit the model
29   model = LinearRegression()
30   model.fit(X_train, y_train)
31
32   # Predicting the Test set results
33   y_pred = model.predict(X_test)
34
35   # Calculating MSE and R-squared
36   mse = mean_squared_error(y_test, y_pred)
37   r_squared = r2_score(y_test, y_pred)
38
39   print(f"Mean Squared Error: {mse}")
40   print(f"R-Squared: {r_squared}")
```

Output Mean Squared Error: 0.245222733362597 R-Squared: 0.6280065468471038

How to Interpret MSE: The smaller the MSE, the closer the model's predictions are to the actual values, which means your model is more accurate in its predictions. An MSE of 0.245 suggests that, on average, the model's predictions deviate from the actual numbers by a relatively small margin — this is a good sign if the scale of the target values is relatively large. However, the acceptability of the MSE value can depend on the context and the range of your data values.

How to Interpret R^2: An R^2 of 0.628 means that about 62.8% of the variance in the dependent variable is predictable from the independent variables. This is moderately good and suggests that the model explains more than half of the variability in the outcome data, but there's still room for improvement.

Visualizing the Linear Regression Model

Understanding the performance of a regression model is greatly enhanced through visualization. The Python code below uses `matplotlib`, a powerful plotting library, to create a visual representation of the training and testing results of a linear regression model. This visualization helps in assessing how well the model has learned the underlying pattern of the data.

Components of the Visualization:

```python
plt.scatter(X_train, y_train, color='blue',
    label='Training data')
plt.scatter(X_test, y_test, color='red',
    label='Test data')
plt.plot(X_test, y_pred, color='black',
    linewidth=2, label='Model fit')

plt.xlabel('X')
plt.ylabel('Y')
plt.title('Linear Regression Fit')
plt.legend()

plt.show()
```

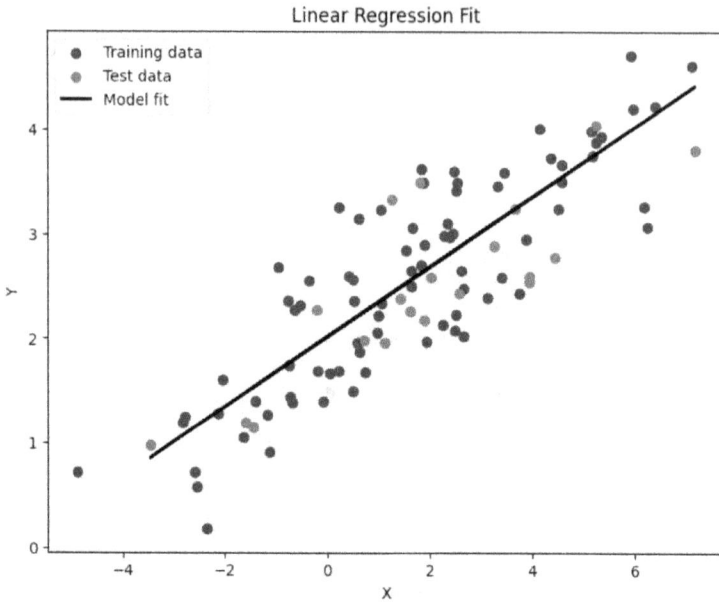

Figure 58. Scatter Plot of Data and Fit

1. **Scatter Plots**:

   - **Training Data**: The blue points represent the training
     data (X_train, y_train). This set is used by the model to
     learn the relationship between the input variable X and
     the output variable Y.
   - **Test Data**: The red points show the test data (X_test,
     y_test). These are new, unseen data points, not used
     in the training phase, that help evaluate how well the
     model generalizes to new data.

2. **Regression Line**:

   - **Model Fit**: The black line plots the model's predictions
     (y_pred) against the test inputs (X_test). This line rep-
     resents the regression equation fitted by the model and

shows how the model predicts the dependent variable Y based on the independent variable X.

Code Explanation:

- `plt.scatter()`: This function creates scatter plots. We use it twice to plot the training data in blue and the test data in red, helping differentiate the data used for learning from the data used for validation.
- `plt.plot()`: This function is used to draw the line representing the model's predictions. The line's properties (color and width) are set to ensure it stands out against the scatter plots, highlighting the model's fit.
- **Axes Labels and Title:**

  - `plt.xlabel('X')` and `plt.ylabel('Y')`: These functions label the x-axis and y-axis, respectively, indicating what each axis represents.
  - `plt.title('Linear Regression Fit')`: Sets the title of the visualization, summarizing what the plot represents.

- `plt.legend()`: Adds a legend to the plot, which is crucial for helping the viewer distinguish between the plotted data points and the regression line.
- `plt.show()`: This function displays the plot. In a script or an interactive session, this call is what actually renders the plot.

This visualization is a key tool in the data scientist's toolkit, providing an immediate visual understanding of how well the model fits the data. It allows for quick identification of patterns, anomalies, or potential areas of improvement in the model, such as underfitting or overfitting. By comparing the scatter of actual data points with the regression line, one can intuitively grasp the effectiveness of the model in capturing the relationship between the variables.

Summary

This example demonstrates the use of MSE and R^2 to evaluate a regression model, providing a clear measure of model accuracy and the variance explained by the model. These metrics are essential for diagnosing regression models and improving model predictions.

Conclusion

Understanding and correctly applying evaluation metrics are fundamental in developing effective regression models. MSE and R^2 offer different perspectives on model performance: MSE provides a measure of error magnitude, while R^2 explains variability in terms of the dependent variable captured by the model. Using these metrics together gives a comprehensive view of model effectiveness.

Overfitting and Underfitting in Machine Learning Models

Understanding overfitting and underfitting is crucial for developing effective machine learning models. These two concepts represent common problems that can significantly impact the performance of a model, both during training and when making predictions on new data.

Overfitting

Overfitting occurs when a model learns the detail and noise in the training data to an extent that it negatively impacts the performance of the model on new data. This means the model is too complex, having too many parameters relative to the number of observations. Overfitted models have learned both the underlying patterns and the random fluctuations in the training data. As a result, they may perform exceptionally well on the training set but poorly on unseen data.

- Characteristics of Overfitting:

    - High accuracy on training data.
    - Poor generalization to new data.

Underfitting

Underfitting occurs when a model is too simple, unable to capture the underlying pattern of the data and, thus, misses the trends in the data. This usually happens when the model does not have enough parameters (or complexity) relative to the scope of the task. Underfitted models fail to achieve a desirable level of performance both on the training data and on new data.

- Characteristics of Underfitting:

    - Poor performance on the training data.
    - Inability to generalize to new data.

Python Code for Visualizing Underfitting and Overfitting

This example will demonstrate underfitting with a simple linear model and overfitting with a high-degree polynomial model. We will plot both to visualize how well they fit the data.

```
1   import numpy as np
2   import matplotlib.pyplot as plt
3   from sklearn.metrics import mean_squared_error
4   from sklearn.model_selection import
5       train_test_split
6   from sklearn.linear_model import LinearRegression
7   from sklearn.preprocessing import
8       PolynomialFeatures
9   from sklearn.pipeline import make_pipeline
10
11  # Generate synthetic data
12  np.random.seed(0)
13  x = np.linspace(-3, 3, 100)
14  y = np.sin(x) + np.random.normal(size=x.shape) *
15      0.3
16  x = x[:, np.newaxis]
17
18  # Splitting dataset into training and testing sets
19  x_train, x_test, y_train, y_test =
20      train_test_split(x, y, test_size=0.3,
21      random_state=42)
22
23  # Function to create a model, plot it, and
24  # calculate MSE
25  def plot_model(degree, x_train, x_test, y_train,
26      y_test, ax, title):
27      model = make_pipeline(PolynomialFeatures
28      (degree), LinearRegression())
29      model.fit(x_train, y_train)
30      y_pred_train = model.predict(x_train)
31      y_pred_test = model.predict(x_test)
32
33      # Plotting
34      ax.scatter(x_train, y_train, color='blue',
35      label='Training data')
```

```
36      ax.scatter(x_test, y_test, color='red',
37      label='Test data')
38      # Sorting x for line plot
39      x_line = np.vstack([x_train, x_test])
40      y_line = model.predict(x_line)
41      sorted_indexes = np.argsort(x_line.flatten())
42      ax.plot(x_line[sorted_indexes].flatten(),
43      y_line[sorted_indexes], color='green',
44      label='Model fit')
45      ax.set_title(title)
46      ax.set_xlabel('X')
47      ax.set_ylabel('Y')
48      ax.legend()
49
50      # Compute MSE for training and testing sets
51      mse_train = mean_squared_error(y_train,
52      y_pred_train)
53      mse_test = mean_squared_error(y_test,
54      y_pred_test)
55      print(f'{title} - Training MSE:
56      {mse_train:.3f}, Test MSE: {mse_test:.3f}')
57
58  # Plotting
59  fig, axs = plt.subplots(1, 2, figsize=(14, 6))
60
61  # Underfitting example with a low degree
62  # polynomial
63  plot_model(1, x_train, x_test, y_train, y_test,
64      axs[0], 'Underfitting Example: Linear Model')
65
66  # Overfitting example with a high degree
67  # polynomial
68  plot_model(15, x_train, x_test, y_train, y_test,
69      axs[1], 'Overfitting Example: High-Degree
70      Polynomial')
```

71
72 `plt.show()`

Underfitting Example: Linear Model - Training MSE: 0.230, Test MSE: 0.318

Overfitting Example: High-Degree Polynomial - Training MSE: 0.064, Test MSE: 0.244

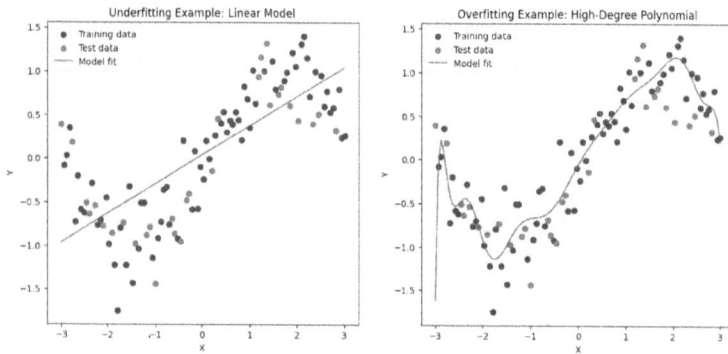

Figure 59. Underfitting and Overfitting Data

Explanation of the Code

- **Data Generation:** The data is synthesized with a sinusoidal pattern perturbed by Gaussian noise.
- **Model Training and Prediction:** The `plot_model` function encapsulates model training using polynomial regression with varying degrees, plotting, and MSE computation.
- **Visualization:**

  - The scatter plots for training and testing data provide a visual basis for understanding model performance.
  - The line plot (`Model fit`) shows the predicted model's behavior across the range of X values.

- **Metrics**: Mean Squared Error (MSE) for both training and test datasets is printed to quantify the model's performance.
- **Subplots**: Two subplots show models with different polynomial degrees to compare underfitting and overfitting visually.

Observations

The results from the two models, as illustrated by the Mean Squared Error (MSE) for both training and testing datasets, provide clear insights into the phenomena of underfitting and overfitting. Here's how to interpret each model's performance based on its MSE values:

Underfitting Example: Linear Model

- **Training MSE**: 0.230
- **Test MSE**: 0.318

Interpretation:

- **Training MSE**: The training MSE of 0.230 suggests that the model does not fit the training data very well. While not exceedingly high, this value indicates that the model's simplicity prevents it from capturing all the underlying patterns in the training data.
- **Test MSE**: The test MSE is slightly higher than the training MSE, at 0.318. This increase in MSE when moving from training to testing data is typical of underfit models, which do not have enough complexity to capture the underlying relationships accurately, even in unseen data. The model's performance degrades when exposed to new data, confirming it doesn't generalize well beyond the training set.

Conclusion: The linear model is too simple to capture the complex relationships in the data adequately. It neither learns the training data well nor generalizes effectively, indicating underfitting.

Overfitting Example: High-Degree Polynomial

- **Training MSE**: 0.064
- **Test MSE**: 0.244

Interpretation:

- **Training MSE**: The low training MSE of 0.064 indicates that the model fits the training data very well—perhaps too well. This suggests that the model has likely captured not only the underlying patterns but also the noise and random fluctuations within the training data.
- **Test MSE**: There is a significant jump in MSE to 0.244 on the test set. This sharp increase is a hallmark of overfitting, where the model, while performing excellently on the training data, fails to maintain this performance on new, unseen data. The model's overly complex nature makes it sensitive to the noise in the training data, which does not generalize to other data sets.

Conclusion: The high-degree polynomial model shows excellent learning on the training set but performs poorly on the test set due to its complexity. It overfits the training data, capturing excessive noise, which undermines its ability to generalize.

Bias vs. Variance

The MSE metrics effectively highlight the trade-off between **bias** and **variance** in these two scenarios:

- The **underfitting model** exhibits high bias, simplifying assumptions about the data's form that leads to errors both on training and testing data.
- The **overfitting model** shows high variance, perfectly tuning itself to the training data at the expense of generalizability, leading to significant errors on new data.

Understanding these dynamics is crucial in machine learning for choosing the right model complexity to balance fitting the training data accurately while maintaining the ability to generalize well to new data. Adjustments in model complexity, regularization techniques, or using more representative training data might help mitigate these issues.

Cross-Validation Techniques

Cross-validation is a robust statistical method used to evaluate the performance of machine learning models. It helps ensure that the model is neither overfitting nor underfitting. This technique is essential for assessing how the results of a statistical analysis will generalize to an independent data set. Here's a deeper look at cross-validation and its commonly used techniques.

Overview of Cross-Validation

Cross-validation involves partitioning a dataset into subsets, training the model on some subsets (training set), and validating it on the remaining subsets (validation set). The key objective is to test the model's ability to predict new data that was not used in estimating it, thus highlighting potential problems like overfitting or underfitting and providing insight into how the model will generalize to an independent dataset.

Common Cross-Validation Techniques

1. **K-Fold Cross-Validation**

   - **Description**: The most widely used form of cross-validation. The data set is randomly partitioned into 'k' equal sized subsets. Of the 'k' subsets, a single subset is retained as the validation data for testing the model, and the remaining 'k-1' subsets are used as training data. The cross-validation process is then repeated 'k' times (the folds), with each of the 'k' subsets used exactly once as the validation data.
   - **Usage**: This method is beneficial because it ensures that every observation from the original dataset has the chance of appearing in training and test set. It is generally used when the training dataset is limited in size, as it maximizes the training data available.

2. **Leave-One-Out (LOO)**

   - **Description**: A particular case of k-fold cross-validation where the number of folds equals the number of instances in the dataset. Thus, for 'n' instances in the dataset, LOO cross-validation yields 'n' different training sets and 'n' different tests set. Each learning set is created by taking all the instances except one, the test set being the single instance left out.
   - **Usage**: LOO is an exhaustive cross-validation technique that provides a comprehensive way of measuring the error of the model but can be very computationally expensive for larger datasets.

3. **Stratified K-Fold Cross-Validation**

   - **Description**: Variation of k-fold which returns stratified folds: each set contains approximately the same percentage of samples of each target class as the complete

set. This is important in classification problems where class imbalance could lead to misrepresentative training or validation sets under standard k-fold validation.
- **Usage**: Particularly useful for imbalanced datasets to ensure that each fold is a good representative of the whole.

4. **Time Series Cross-Validation**

- **Description**: A variant of cross-validation useful in time-series forecasting. Unlike standard cross-validation methods, split points are selected in a way that respects the temporal order of observations.
- **Usage**: Essential for models that predict time-dependent data, where train-test contamination is a risk if random splits are used.

Example in Python: K-Fold Cross-Validation

Here's how you might implement k-fold cross-validation using Python's Scikit-Learn library:

```
from sklearn.model_selection import KFold,
    cross_val_score
from sklearn.ensemble import
    RandomForestClassifier
from sklearn.datasets import load_iris
import numpy as np

# Load Iris dataset
data = load_iris()
X = data.data
y = data.target

```

```
13   # Define the k-fold cross-validator (k=5)
14   kf = KFold(n_splits=5, random_state=None,
15       shuffle=True)
16
17   # Initialize the model
18   model = RandomForestClassifier(n_estimators=100,
19       random_state=42)
20
21   # Perform cross-validation
22   scores = cross_val_score(model, X, y, cv=kf)
23
24   print("Scores from each Iteration: ", scores)
25   print("Average K-Fold Score :", np.mean(scores))
```

Output Scores from each Iteration: [0.93333333 0.9 0.96666667 0.93333333 0.96666667] Average K-Fold Score : 0.9400000000000001

The results you've provided come from a k-fold cross-validation of a RandomForest model on a dataset (likely the Iris dataset, based on common tutorial outcomes). Here's what the results tell us about the model's performance:

Breakdown of the Results

- These scores represent the accuracy of the model for each of the 5 folds used in the cross-validation.
- Each score is the result of the model trained on four of the folds and tested on the remaining fifth fold. This is repeated such that each fold serves as the test set exactly once.
- **Average K-Fold Score:** 0.9400000000000001

    - This average score is calculated from the individual fold scores and represents the overall performance of the model across all folds.

– An average score of approximately 0.94 (or 94%) indicates a high level of accuracy in the model's predictions across the different subsets of data.

Interpretation

- **High Accuracy**: The model achieves high accuracy in each fold, with scores ranging from 90% to 96.67%. Such high results suggest that the RandomForest classifier is very effective for this dataset.
- **Consistency**: The small variation in scores between different folds indicates that the model is stable and performs consistently across different subsets of the data. This consistency is crucial for validating the reliability of the model in practical applications.
- **Generalizability**: An average accuracy of 94% in a 5-fold cross-validation is an excellent result and suggests that the model is not only fitting the training data well but also generalizing effectively to unseen data. This is important for machine learning models, as the ultimate goal is to perform well on new, out-of-sample data.

Conclusion

Overall, the results from this k-fold cross-validation are very positive, demonstrating both high accuracy and good generalizability of the RandomForest model on the dataset in question. The high accuracy across all folds reassures us that the model is not overfitting to a particular subset of the data and is likely to perform well on similar data outside of this dataset.

This type of evaluation is essential for building confidence in the model's predictions and can guide further development and tuning.

For instance, if any folds had significantly lower accuracy, it might indicate areas where the model could be improved or suggest the presence of outliers or anomalies in the data that require further investigation.

Reflective Questions

Here are **10 reflective questions** based on the chapter on **Evaluating Machine Learning Models**:

General Reflection on Model Evaluation

1. Why is it important to evaluate machine learning models beyond just their accuracy?
2. How can imbalanced datasets impact the interpretation of accuracy?

Metrics-Specific Reflections

3. In what situations would you prioritize precision over recall, or vice versa?
4. How does the F1 Score help balance precision and recall, and when is it most useful?
5. What insights can be gained from analyzing False Positives and False Negatives?

Overfitting and Underfitting Reflections

6. How can overfitting be detected when evaluating a model?
7. Why might a model that performs well on training data fail on test data?

Cross-Validation and Model Robustness

8. Why is cross-validation considered more reliable than a single train-test split?
9. How does k-fold cross-validation help assess model generalizability?

Final Reflection on Real-World Use Cases

10. How do you balance simplicity (e.g., accuracy) with more nuanced metrics (e.g., F1 Score) when presenting results to non-technical stakeholders?

Python Challenges

Challenge 1: Calculate All Metrics

Given the following predictions and actual labels, calculate accuracy, precision, recall, F1 Score, and plot the confusion matrix.

```
1   y_true = [1, 0, 1, 1, 0, 1, 0, 1, 0, 0]  # Actual
2       labels
3   y_pred = [1, 0, 1, 0, 0, 1, 0, 1, 1, 0]  #
4       Predicted labels
5
6   from sklearn.metrics import accuracy_score,
7       precision_score, recall_score, f1_score,
8       confusion_matrix
9   import seaborn as sns
10  import matplotlib.pyplot as plt
11
12  # Calculate metrics
13  # <Calculate Accuracy, Precision, Recall,
14  #   and F1 scores here.>
15
16  # Print metrics
17  print(f"Accuracy: {accuracy:.2f}")
18  print(f"Precision: {precision:.2f}")
19  print(f"Recall: {recall:.2f}")
20  print(f"F1 Score: {f1:.2f}")
21
22  # Confusion matrix
23  cm = confusion_matrix(y_true, y_pred)
24  plt.figure(figsize=(6, 4))
25  sns.heatmap(cm, annot=True, fmt="d",
26      cmap="Blues", cbar=False)
27  plt.xlabel("Predicted")
28  plt.ylabel("Actual")
29  plt.title("Confusion Matrix")
30  plt.show()
```

Challenge 2: Adjust Threshold for Better Precision

Given the same dataset, adjust the decision threshold to maximize precision and recalculate the metrics.

```
1   import numpy as np
2
3   # Predicted probabilities
4   y_prob = [0.9, 0.3, 0.8, 0.2, 0.1, 0.85, 0.4, 0.9
5       , 0.6, 0.05]
6
7   # Adjust threshold
8   threshold = 0.7
9   y_pred_adjusted = # <using y_prob, create a new
10                     # array and assign 1 if
11                     # probability
12                     # is above .7, otherwise assign
13                     # 0>
14
15  # Recalculate metrics
16  # <calculate precision, recall, and f1 score
17  #  of y_pred_adjusted here.>
18
19  print(f"Adjusted Precision:
20      {precision_adjusted:.2f}")
21  print(f"Adjusted Recall: {recall_adjusted:.2f}")
22  print(f"Adjusted F1 Score: {f1_adjusted:.2f}")
```

Challenge 3: K-Fold Cross-Validation

Train a decision tree classifier using 5-fold cross-validation on the Iris dataset and report the average precision.

```
1   from sklearn.model_selection import
2       cross_val_score
3   from sklearn.tree import DecisionTreeClassifier
4   from sklearn.datasets import load_iris
5
6   # Load dataset
7   iris = load_iris()
8   X, y = iris.data, iris.target
9
10  # Initialize model
11  model = DecisionTreeClassifier()
12
13  # Perform 5-fold cross-validation
14  scores = # <hint: use cross_val_score with 5
15      folds
16          #for the cv parameter>
17
18  print("Precision Scores for Each Fold:", scores)
19  print("Average Precision:", scores.mean())
```

Challenge 4: Visualizing Overfitting vs. Underfitting

Generate synthetic data and compare the performance of a linear model and a high-degree polynomial model using MSE on training and testing sets.

```python
1   from sklearn.preprocessing import
2       PolynomialFeatures
3   from sklearn.linear_model import LinearRegression
4   from sklearn.pipeline import make_pipeline
5   from sklearn.metrics import mean_squared_error
6   from sklearn.model_selection import
7       train_test_split
8
9   # Generate synthetic data
10  np.random.seed(42)
11  X = np.linspace(-3, 3, 100).reshape(-1, 1)
12  y = np.sin(X).ravel() + np.random.normal
13      (scale=0.2, size=X.shape[0])
14
15  # Split data
16  X_train, X_test, y_train, y_test =
17      train_test_split(X, y, test_size=0.3,
18      random_state=42)
19
20  # Linear model
21  linear_model = make_pipeline(PolynomialFeatures(1
22      ), LinearRegression())
23  linear_model.fit(X_train, y_train)
24  y_train_pred_linear = linear_model.predict
25      (X_train)
26
27  # <Run the prediction on the test data here.>
28  y_test_pred_linear = #<fit the test data
29                       # on the linear model here>
30
31  # Polynomial model
32  poly_model = make_pipeline(PolynomialFeatures(15)
33      , LinearRegression())
34  poly_model.fit(X_train, y_train)
35  y_train_pred_poly = poly_model.predict(X_train)
```

```
36
37   y_test_pred_poly = #<fit the test data
38                      # on the polynomial model
39   # here>
40
41   # Calculate MSE
42   # <Use the mean_squared_error function to predict
43   # <linear training data error>
44   # <linear test data error>
45   # <polynomial training data error>
46   # <and the polynomial test data error>
47   mse_train_linear = #<calculate error here>
48   mse_test_linear =  #<calculate error here>
49   mse_train_poly = #<calculate error here>
50   mse_test_poly = #<calculate error here>
51
52   print(f"Linear Model - Train MSE:
53       {mse_train_linear:.2f}, Test MSE:
54       {mse_test_linear:.2f}")
55   print(f"Polynomial Model - Train MSE:
56       {mse_train_poly:.2f}, Test MSE:
57       {mse_test_poly:.2f}")
58
59   # Plot results
60   plt.figure(figsize=(12, 6))
61   plt.scatter(X_train, y_train, color="blue",
62       label="Training data")
63   plt.scatter(X_test, y_test, color="red",
64       label="Testing data")
65   plt.plot(X, linear_model.predict(X),
66       color="green", label="Linear model")
67   plt.plot(X, poly_model.predict(X), color="orange"
68       , label="Polynomial model")
69   plt.legend()
70   plt.title("Overfitting vs. Underfitting")
```

```
71   plt.show()
```

_____

Chapter 13: Hyperparameter Tuning

Hyperparameter tuning is a crucial step in the process of designing and training machine learning models. Hyperparameters, unlike model parameters, are set before the learning process begins and directly influence the behavior and performance of the models built. This chapter begins with discussing the importance of tuning hyperparameters and how it affects the overall outcomes of machine learning projects.

Importance of Tuning Hyperparameters

Hyperparameters are the settings that can be configured prior to training a model and are not directly learned from the data. Examples include the learning rate, number of trees in a random forest, the depth of trees, or the number of hidden layers and neurons in a neural network. Proper tuning of these hyperparameters plays a vital role in controlling the model training process and has a significant impact on the performance of the model.

1. Enhancing Model Performance:
The primary reason to tune hyperparameters is to optimize the model's performance. The default settings of hyperparameters are seldom ideal for all problems. Each dataset has its characteristics, and fine-tuning the hyperparameters to align with these specifics can drastically improve how well the model learns and generalizes.

2. Preventing Overfitting and Underfitting:
Hyperparameters can control the model's complexity. For instance, too many trees in a random forest or too deep trees can lead to overfitting where the model learns the training data too well, including the noise and outliers, failing to generalize on unseen data. Conversely, overly simplistic models might underperform (underfit) due to insufficient learning capacity, unable to capture the underlying pattern of the data. Hyperparameter tuning helps find the sweet spot to balance between these two extremes.

3. Improving Learning Efficiency:
Some hyperparameters determine the efficiency of the learning process, such as the learning rate in gradient descent. A too high learning rate can cause the learning process to converge too quickly to a suboptimal solution, while a too low rate can slow down the process, resulting in long training times without substantial gains in performance. Tuning these can lead to more efficient learning processes, saving both time and computational resources.

4. Model Adaptability:
Different datasets may benefit from different learning algorithms and settings. Hyperparameter tuning allows models to adapt better to new data or different types of data, enhancing the model's utility across various tasks in an adaptable and robust manner.

Practical Approach to Hyperparameter Tuning

The process of hyperparameter tuning can be approached in several ways:

- **Grid Search**: Testing a specific set of hyperparameter values. It is thorough but can be computationally expensive.

- **Random Search**: Randomly testing a wide range of hyper-parameter values. This method can sometimes find good configurations more quickly than grid search.
- **Automated Hyperparameter Optimization Techniques**: Methods like Bayesian Optimization, Gradient-based optimization, and Evolutionary Algorithms, which aim to find the best hyperparameters in a more systematic and less resource-intensive manner.

Each of these methods has its advantages and scenarios where it is best used. The subsequent sections of this chapter will dive deeper into these strategies, providing examples and case studies to illustrate effective hyperparameter tuning practices in machine learning projects.

This foundational understanding sets the stage for exploring specific hyperparameter tuning techniques and tools that can help achieve the best model performance across various machine learning tasks.

Grid Search and Random Search

When training machine learning models, selecting the optimal combination of hyperparameters can dramatically affect performance. Two of the most popular methods for hyperparameter tuning are Grid Search and Random Search. Each offers a different approach to exploring the hyperparameter space, and understanding their strengths and weaknesses is essential for effectively deploying them.

Grid Search

Grid Search is a systematic way of thoroughly testing and tuning hyperparameters by setting up a grid of hyperparameter values and evaluating their performance.

How It Works:

1. **Define the Grid**: Specify a grid of hyperparameter values. This grid represents all combinations of hyperparameter values to be tested.
2. **Exhaustive Testing**: Train a model on each combination of the specified hyperparameter values in the grid. Typically, this is done using cross-validation to ensure the model's performance is robust and not dependent on a particular split of the training data.
3. **Select the Best Model**: After training models on all the combinations, select the model with the best performance metric, typically accuracy, F1 score, or Mean Squared Error (MSE), depending on the problem.

Pros:

- Comprehensive: By exhaustively searching through the specified range, grid search can find the best combination of hyperparameters from the set provided.
- Easy to Understand: The process is straightforward and easy to implement with libraries such as Scikit-Learn.

Cons:

- Computationally Expensive: As the number of hyperparameters grows, the grid grows exponentially (known as the "curse of dimensionality"), making grid search computationally expensive and sometimes impractical.
- Limited to Discrete Values: Grid search only explores the hyperparameters at the grid points, potentially missing the optimal settings in between.

Python Example Using Scikit-Learn:

```
1   from sklearn.model_selection import GridSearchCV
2   from sklearn.ensemble import
3       RandomForestClassifier
4   from sklearn.datasets import load_iris
5
6   # Load data
7   data = load_iris()
8   X, y = data.data, data.target
9
10  # Set up the hyperparameter grid
11  # 3 x 2 x 5 only leaves us with 30 combinations
12  param_grid = {
13      'n_estimators': [50, 100, 200],
14      'max_features': ['sqrt', 'log2'],
15      'max_depth' : [4, 5, 6, 7, 8]
16  }
17
18  # Create a base model
19  rf = RandomForestClassifier()
20
21  # Instantiate the grid search model
22  grid_search = GridSearchCV(
23      estimator = rf,
24      param_grid = param_grid,
25      cv = 3, n_jobs = -1,
26      verbose = 2)
27
28  # Fit the grid search to the data
29  grid_search.fit(X, y)
30  print("Best parameters found: ",
31      grid_search.best_params_)
```

Output Fitting 3 folds for each of 30 candidates, totalling 90 fits
Best parameters found: {'max_depth': 4, 'max_features': 'sqrt',
'n_estimators': 100}

The Grid Search was performed with 3-fold cross-validation on a combination of 30 different sets of hyperparameters, resulting in a total of 90 model fits. The best performing model used the parameters {'max_depth': 4, 'max_features': 'sqrt', 'n_estimators': 100}, meaning it found that a maximum depth of 4, using the square root of the number of features at each split, and 100 trees, gave the optimal performance for the RandomForest based on the criteria set in the grid search configuration.

Random Search

Random Search differs from grid search in that it samples hyperparameter settings randomly from a specified distribution for a fixed number of iterations. This approach can outperform grid search when only a few hyperparameters actually influence the final performance of the machine learning model.

How It Works:

1. **Define the Distribution**: Specify a distribution or range of values for each hyperparameter from which values can be randomly sampled.
2. **Random Sampling**: Randomly sample combinations of hyperparameters from the specified distributions.
3. **Train and Evaluate**: Like grid search, train a model on each combination and evaluate using cross-validation.
4. **Select the Best Model**: Choose the hyperparameters that result in the best performance.

Pros:

- Efficient: Can search a larger space with fewer trials because it does not need to try every combination but rather selects them randomly.
- Effective in High Dimensions: Better suited than grid search when dealing with a large number of hyperparameters.

Cons:

- Less Comprehensive: There's no guarantee of finding the optimal solution since it's based on random sampling.
- Requires More Trials: To get close to the best hyperparameters, a large number of iterations might be necessary.

Python Example Using Scikit-Learn:

```
from sklearn.model_selection import
    RandomizedSearchCV
from sklearn.ensemble import
    RandomForestClassifier
from sklearn.datasets import load_iris
import scipy.stats as stats

# Load data
data = load_iris()
X, y = data.data, data.target

# Hyperparameter distribution
param_dist = {
    'n_estimators': stats.randint(10, 200),
    'max_features': ['sqrt', 'log2'],
    'max_depth': stats.randint(3, 10)
}

# Create a base model
rf = RandomForestClassifier()

# Instantiate the random search model
random_search = RandomizedSearchCV(
    estimator = rf,
    param_distributions = param_dist,
    n_iter = 100,
```

```
27        cv = 3, verbose=2, random_state=42,
28        n_jobs = -1)
29
30    # Fit the random search model
31    random_search.fit(X, y)
32    print("Best parameters found: ",
33        random_search.best_params_)
```

Output Fitting 3 folds for each of 100 candidates, totalling 300 fits Best parameters found: {'max_depth': 9, 'max_features': 'log2', 'n_estimators': 102}

The results from the random search indicate that it was performed with 3-fold cross-validation across 100 different sets of hyperparameters, totaling 300 model fits. The optimal parameters found were {'max_depth': 9, 'max_features': 'log2', 'n_estimators': 102}. This suggests a more complex model with a greater maximum depth of 9, using logarithm base 2 of the number of features at each split, and slightly more trees (102) compared to the grid search result.

Tuning Comparison

You can directly compare the performance of models obtained from Grid Search and Random Search on the training data by calculating various metrics such as accuracy, precision, recall, F1 score, or even the loss metrics like Mean Squared Error (MSE) depending on the type of problem (classification or regression). Here's how you can do it:

```
1    # Predicting on training data
2    y_train_pred_grid = grid_search.predict(X)
3    y_train_pred_random = random_search.predict(X)
4
5    # Evaluating accuracy
6    accuracy_grid = accuracy_score(y,
7        y_train_pred_grid)
8    accuracy_random = accuracy_score(y,
9        y_train_pred_random)
10
11   print(f"Training Accuracy from Grid Search:
12       {accuracy_grid}")
13   print(f"Training Accuracy from Random Search:
14       {accuracy_random}")
```

Output Training Accuracy from Grid Search: 0.9866666666666667
Training Accuracy from Random Search: 1.0

The results show that the model trained with hyperparameters obtained from Random Search achieved perfect training accuracy (1.0), indicating that it has learned to predict every training instance correctly. In contrast, the model optimized via Grid Search also performed exceptionally well on the training data, achieving a training accuracy of approximately 98.67%, but slightly less than the Random Search model, potentially indicating less overfitting to the training data.

Interpretation

- **Higher Accuracy**: If one model shows significantly higher accuracy on the training data, it might be better tuned for the particular characteristics of the training set.
- **Consider Overfitting**: Exceptionally high training accuracy as compared to validation or test accuracy can indicate overfitting. It's crucial to ensure that improvements

in training performance correspond to improvements in test performance.

This comparison can provide valuable insights into how each tuning method has impacted the model's ability to learn from the training data. It is also advisable to monitor performance on a separate validation or test set to ensure that the model remains generalizable to new data.

Conclusion

Both grid search and random search are powerful techniques for hyperparameter tuning, each with its own advantages and trade-offs. Choosing between them depends on the specific needs of the project, the computational resources available, and the number of hyperparameters to tune. In practice, it's common to start with random search to narrow down the range and then fine-tune the model with grid search within the identified ranges.

For the last section of this chapter, we'll dive into a hands-on activity focused on tuning hyperparameters for a Random Forest model using the Kaggle *Wine Quality* dataset. This process will demonstrate how adjustments to hyperparameters can enhance model performance. Tuning hyperparameters is essential in machine learning, especially with ensemble methods like Random Forests, which have several configurable parameters that can significantly impact accuracy and generalization.

Hands-on Activity: Hyperparameter Tuning with Random Forests on the Wine Quality Dataset

In this activity, we'll perform the following steps:

1. **Load and Explore the Dataset**: The Wine Quality dataset contains information about various chemical properties of wine and their relationship with quality ratings.
2. **Prepare the Data**: Handle any missing values and split the data into training and testing sets.
3. **Train a Baseline Model**: Use a default Random Forest model to establish baseline performance.
4. **Tune Hyperparameters Using Grid Search**: Explore different hyperparameters to improve model performance.
5. **Evaluate the Tuned Model**: Compare the baseline and tuned model to see if performance improves.

Step 1: Load and Explore the Dataset

```
1    import pandas as pd
2
3    # Load the dataset
4    data = pd.read_csv('wine_dataset.csv')
5
6    # Preview the dataset
7    data.head()
```

Output

	fixed_acidity	volatile_acidity	citric_acid	residual_sugar	chlorides	free_sulfur_dioxide	total_sulfur_dioxide	density	pH	sulphates	alcohol	quality	style
0	7.4	0.70	0.00	1.9	0.076	11.0	34.0	0.9978	3.51	0.56	9.4	5	red
1	7.8	0.88	0.00	2.6	0.098	25.0	67.0	0.9968	3.20	0.68	9.8	5	red
2	7.8	0.76	0.04	2.3	0.092	15.0	54.0	0.9970	3.26	0.65	9.8	5	red
3	11.2	0.28	0.56	1.9	0.075	17.0	60.0	0.9980	3.16	0.58	9.8	6	red
4	7.4	0.70	0.00	1.9	0.076	11.0	34.0	0.9978	3.51	0.56	9.4	5	red

Figure 60. head of the wine dataset

The dataset contains various chemical properties like acidity, sugar content, pH, and alcohol, along with a quality score for each wine sample.

Below is general info about the dataset structure:

```
1  # dataset info
2  data.info()
```

Output

```
1  RangeIndex: 6497 entries, 0 to 6496
2  Data columns (total 13 columns):
3  No.  Column                Non-Null Count  Dtype
4  ___  _____                _____  _____
5  0    fixed_acidity         6497 non-null   float64
6  1    volatile_acidity      6497 non-null   float64
7  2    citric_acid           6497 non-null   float64
8  3    residual_sugar        6497 non-null   float64
9  4    chlorides             6497 non-null   float64
10   5    free_sulfur_dioxide   6497 non-null   float64
11   6    total_sulfur_dioxide  6497 non-null   float64
12   7    density               6497 non-null   float64
13   8    pH                    6497 non-null   float64
14   9    sulphates             6497 non-null   float64
15   10   alcohol               6497 non-null   float64
16   11   quality               6497 non-null   int64
17   12   style                 6497 non-null   object
18   dtypes: float64(11), int64(1), object(1)
19   memory usage: 660.0+ KB
```

The Wine Quality dataset consists of 6,497 entries with 13 columns, capturing various chemical properties and characteristics of wine samples. Each column is complete, with no missing values, suggesting a well-prepared dataset for analysis. There are 11 numerical columns, primarily containing float values for the chemical properties, one integer column representing wine quality scores, and one categorical column labeled "style," distinguishing between different types of wine, such as red or white. This dataset provides comprehensive information for evaluating how chemical attributes may influence wine quality.

Step 2: Prepare the Data

```
1  from sklearn.model_selection import
2      train_test_split
3  from sklearn.preprocessing import LabelEncoder
4
5  # Change Style into something the model can work
6  # with
7
8  # Convert the 'style' column using Label Encoding
9  label_encoder = LabelEncoder()
10 data['style'] = label_encoder.fit_transform
11     (data['style'])
12
13 # Verify the transformation
14 print(data['style'].head())
15
16
17 # Separate features and target variable
18 X = data.drop('quality', axis=1)
19 y = data['quality']
20
21
22
23 # Split data into training and testing sets
24 X_train, X_test, y_train, y_test =
25     train_test_split(X, y, test_size=0.2,
26     random_state=42)
```

In this step, we split the data into training and testing sets with 80% for training and 20% for testing.

Step 3: Train a Baseline Model

We'll first train a Random Forest model with default hyperparameters to see how it performs before tuning.

```
1   from sklearn.ensemble import
2       RandomForestClassifier
3   from sklearn.metrics import accuracy_score
4
5   # Initialize the Random Forest model
6   rf = RandomForestClassifier(random_state=42)
7
8   # Fit the model to the training data
9   rf.fit(X_train, y_train)
10
11  # Make predictions
12  y_pred = rf.predict(X_test)
13
14  # Evaluate the model
15  baseline_accuracy = accuracy_score(y_test, y_pred)
16  print(f"Baseline Accuracy:
17      {baseline_accuracy:.2f}")
```

Output Baseline Accuracy: 0.66

This baseline accuracy will serve as our reference point to see if hyperparameter tuning improves the model's performance.

Step 4: Tune Hyperparameters Using Grid Search

We'll use a Grid Search to test different combinations of hyperparameters. In a Random Forest, common hyperparameters include n_estimators (number of trees), max_depth (maximum depth of each tree), and min_samples_split (minimum samples needed to split a node).

```
1    from sklearn.model_selection import GridSearchCV
2
3    # Define the hyperparameters and their values for
4    # the grid search
5    param_grid = {
6        'n_estimators': [50, 100, 200],
7        'max_depth': [None, 10, 20, 30],
8        'min_samples_split': [2, 5, 10],
9        'min_samples_leaf': [1, 2, 4]
10   }
11
12   # Initialize the Grid Search with cross-validation
13   grid_search = GridSearchCV
14       (estimator=RandomForestClassifier
15                       (random_state=42),
16       param_grid=param_grid,
17       cv=3,
18       n_jobs=-1,
19       verbose=2)
20
21   # Fit the grid search to the training data
22   grid_search.fit(X_train, y_train)
23
24   # Get the best model and parameters
25   best_rf = grid_search.best_estimator_
26   print(f"Best Parameters:
27       {grid_search.best_params_}")
```

Output Fitting 3 folds for each of 108 candidates, totalling 324 fits
Best Parameters: {'max_depth': 20, 'min_samples_leaf': 1, 'min_-
samples_split': 2, 'n_estimators': 200}

Here, we use cross-validation with cv=3 to avoid overfitting and to
ensure that our model's performance is consistent across different
folds of the training data. The GridSearchCV will automatically
try all combinations of specified hyperparameters and return the

model with the best results.

Step 5: Evaluate the Tuned Model

Now that we have the best model, we can test its performance on the test data to see if tuning has led to any improvement.

```
1   # Make predictions with the best model
2   y_pred_tuned = best_rf.predict(X_test)
3
4   # Evaluate the tuned model
5   tuned_accuracy = accuracy_score(y_test,
6       y_pred_tuned)
7   print(f"Tuned Model Accuracy:
8       {tuned_accuracy:.2f}")
9
10  # Compare baseline and tuned model accuracies
11  improvement = ((tuned_accuracy -
12      baseline_accuracy) / baseline_accuracy) * 100
13  print(f"Accuracy Improvement: {improvement:.2f}%")
```

Output Tuned Model Accuracy: 0.68 Accuracy Improvement: 2.31%

The results show that our hyperparameter tuning led to a tuned model accuracy of 68%, reflecting a modest improvement of 2.31% over the baseline model. This comparison helps us understand the effectiveness of our tuning process, even if the improvement was relatively small. Hyperparameter tuning often enhances model performance, but the extent of improvement depends on factors like data quality, feature selection, and the specific hyperparameter combinations explored. In this case, the tuning yielded incremental gains, suggesting the model may have reached a performance plateau with the current dataset.

Summary

In this hands-on activity, we tuned a Random Forest model using Grid Search to find the best hyperparameters for the *Wine Quality* dataset. Hyperparameter tuning helps optimize model performance and can yield substantial improvements, as seen by comparing the baseline and tuned models. This process illustrates the importance of tuning and testing different parameter values to achieve the best results for predictive tasks.

As an exercise, apply **Random Search** on the Wine Quality dataset to explore different configurations, compare the results, and see if this alternative tuning method enhances the model's performance further.

_____

Below are some questions to reflect on to help enforce your learning on Hyperparameter Tuning:

Reflective Questions

1. **Conceptual Questions:**

    - What are hyperparameters, and how are they different from model parameters?
    - Why is hyperparameter tuning crucial for improving machine learning model performance?
    - How do hyperparameters influence overfitting and underfitting in a model?
    - Compare and contrast Grid Search and Random Search. What are the advantages and disadvantages of each?
    - Why might Random Search outperform Grid Search in certain scenarios?

- What role does cross-validation play in hyperparameter tuning?
- Explain how hyperparameter tuning impacts the efficiency of a learning process.

2. **Practical Questions**:

- Describe the steps involved in performing a Grid Search for hyperparameter tuning.
- What are the most common hyperparameters tuned in a Random Forest model?
- How would you interpret results from Grid Search when two or more combinations yield similar performance metrics?
- How do you avoid overfitting during hyperparameter tuning?
- How can you evaluate the success of a tuned model compared to a baseline model?

3. **Advanced Questions**:

- What are some automated hyperparameter optimization techniques, and how do they differ from Grid Search and Random Search?
- Discuss the impact of computational cost when performing hyperparameter tuning on large datasets.
- How would you prioritize which hyperparameters to tune when faced with limited resources?

Python Challenges for Chapter 13: Hyperparameter Tuning

Challenge 1: K-Nearest Neighbors Hyperparameter Tuning

Problem: Use GridSearchCV to tune the hyperparameters of a K-Nearest Neighbors (KNN) classifier on the Iris dataset. Tune the n_neighbors and weights parameters. Report the best hyperparameters and the test set accuracy.

```
1  from sklearn.model_selection import GridSearchCV,
2      train_test_split
3  from sklearn.neighbors import KNeighborsClassifier
4  from sklearn.datasets import load_iris
5  from sklearn.metrics import accuracy_score
6
7  # Load the dataset
8  data = load_iris()
9  X, y = data.data, data.target
10
11 # Split the data
12 X_train, X_test, y_train, y_test =
13     train_test_split(X, y, test_size=0.2,
14     random_state=42)
15
16 # Define the parameter grid
17 param_grid = {
18     'n_neighbors': [3, 5, 7, 9],
19     'weights': ['uniform', 'distance']
20 }
21
22 # Initialize GridSearchCV with KNN
23 # <Perform your grid search here>
```

```
24
25   # Fit the model
26   # <Fit he grid search model here>
27
28   # Best parameters and test accuracy
29   best_params = grid_search.best_params_
30   y_pred = grid_search.best_estimator_.predict
31       (X_test)
32   accuracy = accuracy_score(y_test, y_pred)
33
34   print(f"Best Parameters: {best_params}")
35   print(f"Test Set Accuracy: {accuracy:.2f}")
```

Challenge 2: Random Search on Decision Tree Classifier

Problem: Perform a Random Search on the Wine Quality dataset to optimize the hyperparameters of a Decision Tree classifier. Tune the `max_depth`, `min_samples_split`, and `criterion` parameters. Report the best hyperparameters and the test accuracy.

```
1    from sklearn.datasets import load_wine
2    from sklearn.tree import DecisionTreeClassifier
3    from sklearn.model_selection import
4        RandomizedSearchCV, train_test_split
5    from sklearn.metrics import accuracy_score
6    import scipy.stats as stats
7
8    # Load the dataset
9    data = load_wine()
10   X, y = data.data, data.target
```

```
11
12  # Split the data
13  X_train, X_test, y_train, y_test =
14      train_test_split(X, y, test_size=0.2,
15      random_state=42)
16
17  # Define the hyperparameter distribution
18  param_dist = {
19      'max_depth': stats.randint(3, 20),
20      'min_samples_split': stats.randint(2, 20),
21      'criterion': ['gini', 'entropy'],
22  }
23
24  # Initialize RandomizedSearchCV with
25  # DecisionTreeClassifier
26  # <Perform your random search here>
27
28  # Fit the model
29  # <Fit the random search model here>
30
31  # Best parameters and test accuracy
32  best_params = random_search.best_params_
33  y_pred = random_search.best_estimator_.predict
34      (X_test)
35  accuracy = accuracy_score(y_test, y_pred)
36
37  print(f"Best Parameters: {best_params}")
38  print(f"Test Set Accuracy: {accuracy:.2f}")
```

Challenge 3: Visualizing Hyperparameter Search Results

Problem: Use `GridSearchCV` to tune the `max_depth` and `n_estimators` hyperparameters of a Random Forest classifier on the Iris dataset.   Visualize the results of the Grid Search as a heatmap showing the accuracy for each combination of hyperparameters.

```
import matplotlib.pyplot as plt
import seaborn as sns
from sklearn.ensemble import
    RandomForestClassifier

# Load the dataset
data = load_iris()
X, y = data.data, data.target

# Split the data
X_train, X_test, y_train, y_test =
    train_test_split(X, y, test_size=0.2,
    random_state=42)

# Define the parameter grid
param_grid = {
    'max_depth': [3, 5, 7],
    'n_estimators': [50, 100, 150]
}

# Initialize GridSearchCV with
# RandomForestClassifier

# <Create a grid_search cross validation model>
# <using a RandomForestClassifier here.>

```

```
27   # <Fit the grid search model on the training
28   # data here.>
29
30   # Extract results for heatmap
31   results = grid_search.cv_results_
32   heatmap_data = results['mean_test_score'].reshape
33       (len(param_grid['max_depth']), len
34       (param_grid['n_estimators']))
35
36   # Plot heatmap
37   sns.heatmap(heatmap_data, annot=True,
38       xticklabels=param_grid['n_estimators'],
39       yticklabels=param_grid['max_depth'])
40   plt.xlabel('Number of Estimators')
41   plt.ylabel('Max Depth')
42   plt.title('Grid Search Accuracy Heatmap')
43   plt.show()
```

Chapter 14: Ethical Considerations in Data Analytics and ML - Bias

As data analytics and machine learning increasingly influence decision-making in areas like healthcare, criminal justice, finance, and employment, understanding the ethical implications of these technologies is critical. Data-driven insights can empower organizations to make more informed decisions, but they can also inadvertently propagate or even amplify biases present in the data. In this unit, we will explore the ethical challenges in data analytics and machine learning, focusing on identifying, understanding, and mitigating the risks of bias. By addressing these ethical concerns, we can strive to create fairer, more transparent, and more equitable models that serve diverse populations responsibly.

Bias in Data and Algorithms

Bias is an inherent risk in machine learning and data analytics, arising from the complexities of data collection, processing, and model development. When bias is embedded in data or algorithms, it can lead to outcomes that are systematically unfair or inaccurate for certain groups. For example, a predictive model used for hiring may unfairly disadvantage individuals from specific backgrounds if the training data reflects historical prejudices. Recognizing and addressing these biases is crucial to ensure that models are not only accurate but also ethically aligned with the principles of fairness and equality.

In this chapter, we focus on the origins of bias in data and algorithms, examining how different sources of bias can influence the performance and fairness of machine learning models. We'll discuss the types of bias that arise at various stages of the data pipeline and explore how choices made in data collection, feature selection, and model training can affect outcomes. Understanding the sources of bias is the first step in designing strategies to identify and mitigate bias, promoting more equitable machine learning practices.

Sources of Bias in Data

Bias in machine learning models often originates from the data itself, where historical, social, and systemic biases may already be embedded. Understanding these sources of bias is essential for building models that minimize harmful or discriminatory impacts. Below are some of the primary sources of bias in data:

1. **Historical and Societal Biases**

   - Data collected from human activities reflects existing societal structures, norms, and inequalities. If the data reflects historical inequities—such as differences in income, education, or access to services among different groups—models trained on this data may perpetuate these inequities. For instance, a hiring model trained on historical data from a company may disadvantage candidates from minority groups if the company's past hiring practices were biased.

2. Sampling Bias

- Sampling bias occurs when the data used to train a model is not representative of the broader population that the model will serve. This can happen if the dataset is imbalanced, overrepresenting some groups while underrepresenting others. For example, a facial recognition system trained primarily on images of lighter-skinned individuals may perform poorly when recognizing darker-skinned individuals. Sampling bias is especially concerning in applications like healthcare, where unrepresentative training data can lead to inaccurate predictions and recommendations for underserved groups.

3. Measurement Bias

- Measurement bias arises from inconsistencies or inaccuracies in how data is measured, recorded, or labeled. In some cases, certain attributes are either overemphasized or underrepresented due to flawed measurement methods. For instance, using credit score as a proxy for trustworthiness in loan applications may disproportionately disadvantage low-income individuals who have limited credit history, even if they are reliable borrowers. Measurement bias can also occur if there is a systematic error in how data points are collected, such as underreporting of certain crimes in specific communities.

4. Label Bias

- Label bias can be introduced when human judgments or subjective criteria are used to label data. For instance, in datasets for crime prediction models, the labels may reflect societal biases present in the justice system. Arrest data, which may be used to label instances of criminal behavior, often reflects disparities in law enforcement practices across different demographic groups. Similarly, sentiment analysis models trained on labeled social media posts can inherit biases if human annotators' personal beliefs influence how they label the sentiments of certain phrases.

5. Observer Bias

- Observer bias occurs when the individuals involved in data collection or labeling introduce their personal biases, consciously or unconsciously. For example, survey data can be biased if interviewers, based on their own beliefs or assumptions, influence respondents' answers. This type of bias is often subtle but can significantly impact the quality of data used in machine learning models.

6. Confirmation Bias in Data Collection

- Confirmation bias occurs when data collectors, consciously or unconsciously, seek information that confirms their existing beliefs or assumptions. For instance, if researchers conducting a study on job performance only gather data from high-performing employees, the resulting dataset may fail to represent the range of performance seen across the full workforce. This leads to models that may inaccurately predict or assess job performance for broader populations.

7. Proxy Variables and Implicit Bias

- Proxy variables are variables that are not directly relevant to the prediction task but are correlated with sensitive attributes, such as race, gender, or socioeconomic status. For example, using ZIP code as a feature in models predicting loan eligibility could introduce implicit bias, as ZIP codes are often correlated with demographic and socioeconomic factors. Models using these proxy variables may inadvertently discriminate against certain groups, even if sensitive attributes are not explicitly included.

8. **Data Imbalance**

- When the distribution of categories in the data is highly imbalanced, models may struggle to learn patterns for underrepresented groups, resulting in biased predictions. For example, in medical diagnosis datasets, conditions that are rare or less frequently recorded may be underrepresented, leading to models that fail to diagnose these conditions accurately. Data imbalance often skews predictions towards the majority class, disadvantaging minority groups in real-world applications.

By identifying the sources of bias in data, practitioners can make informed decisions about how to mitigate these biases during data collection, preprocessing, and model development. In the following sections, we will discuss strategies for detecting and addressing these biases to promote fairness and inclusivity in machine learning models. Recognizing these sources is essential for designing models that do not simply replicate historical or societal inequities but instead work towards more equitable outcomes.

How Machine Learning Models Can Perpetuate Bias

Machine learning models can unintentionally perpetuate and even amplify biases present in the data they are trained on. Because these models learn patterns from historical data, they often inherit and propagate any existing prejudices, disparities, or inequities within that data. This can lead to decisions and recommendations that are skewed against certain groups, impacting individuals and communities in areas like employment, healthcare, criminal justice, and lending.

Understanding how machine learning models can perpetuate bias helps us identify potential issues in model design, data usage, and deployment. Below are some of the main ways that bias in machine learning models can arise and persist, even after training:

1. **Reinforcing Historical Biases**

   - Machine learning models are built to detect patterns and correlations in data, but when the data itself reflects historical biases or social inequities, models may reinforce these biases. For example, if a model is trained on hiring data from a company that historically hired fewer women in leadership roles, the model might favor male candidates in future predictions. Such biases become embedded in the model, creating a feedback loop where biased predictions lead to biased real-world outcomes.

2. **Overfitting to Skewed Data**

   - When a model is trained on a dataset that is unrepresentative of the broader population, it can overfit to patterns that are specific to the skewed data. Overfitting occurs when a model learns very specific patterns in the training data that may not generalize well. For instance, a medical diagnostic model trained primarily on data from one demographic group may struggle to diagnose individuals from other groups accurately. This perpetuates bias by producing accurate predictions only for overrepresented groups, while yielding subpar predictions for underrepresented ones.

3. **Algorithmic Amplification of Bias**

- Machine learning algorithms, especially those that prioritize accuracy or performance, can amplify biases in data. For example, recommendation algorithms often optimize based on user engagement data, which can lead to a reinforcement of existing preferences or stereotypes. In a hiring recommendation system, an algorithm that prioritizes past hiring patterns might amplify biases by suggesting candidates with profiles similar to previously hired individuals, thereby marginalizing candidates from less-represented backgrounds.

4. **Bias from Proxy Variables**

- Models often rely on proxy variables—features that are not explicitly about sensitive attributes but are correlated with them. For instance, in predicting creditworthiness, models may use ZIP codes as a feature. ZIP codes can correlate with socioeconomic factors, race, and access to resources, creating an implicit bias. Although the model does not directly use race as an attribute, it may indirectly discriminate based on ZIP code, perpetuating bias in ways that may be hard to detect.

5. **Label Bias in Supervised Learning**

- In supervised learning, models rely on labeled data to learn associations between inputs and outputs. However, labels are often subject to human judgment and can reflect subjective biases. For instance, sentiment analysis models trained on social media posts may inherit bias if the labels are influenced by annotators' personal beliefs or cultural assumptions. A crime prediction model trained on historical arrest data may embed biases in law enforcement practices, leading to predictions that disproportionately target certain demographic groups.

6. **Bias in Model Evaluation Metrics**

- Even the way models are evaluated can perpetuate bias if metrics are not chosen carefully. Many evaluation metrics, such as accuracy, do not account for fairness and may mask disparities in model performance across different groups. For example, a model that performs well overall might still perform poorly for minority groups if accuracy across all groups isn't evaluated separately. This can result in models that seem accurate but consistently make errors for specific demographics, perpetuating biased outcomes in deployment.

7. **Unintended Consequences from Feedback Loops**

- When models are deployed in real-world systems, they may influence future data collection, creating feedback loops that reinforce biases. For instance, a predictive policing model may lead police to patrol certain neighborhoods more frequently, resulting in more recorded crimes in those areas. This data is then fed back into the model, reinforcing the pattern and creating a self-fulfilling cycle of bias. Similar feedback loops can occur in other domains, such as lending, where biased decisions impact credit scores, influencing future loan decisions and perpetuating inequality.

8. **Lack of Diversity in Model Training and Testing Phases**

- If model developers do not prioritize diversity in both the training and testing stages, the model may fail to account for variability across different groups. For example, a facial recognition model trained and tested primarily on lighter-skinned individuals may perform poorly on darker-skinned faces. This lack of diversity in model development contributes to biased outcomes, which can lead to discriminatory practices in applications such as security and surveillance.

Strategies to Mitigate Bias in Models

Mitigating bias in machine learning models is essential for promoting fairness and preventing harmful or inequitable outcomes. Addressing bias requires a proactive approach that spans the entire machine learning pipeline, from data collection to model evaluation and deployment. Below are several strategies for mitigating bias in models:

1. **Collect Diverse and Representative Data**

- The first step to minimizing bias in models is to ensure that the training data is representative of the population the model is intended to serve. This involves gathering data from diverse sources and balancing the representation of various demographic groups. For example, if a model will be used in healthcare, it should include data from patients across different ages, genders, races, and socioeconomic backgrounds to avoid skewing predictions toward overrepresented groups.

- Regularly updating data is also important, as populations and societal norms evolve over time. Using outdated data may introduce biases that do not reflect current realities.

2. Identify and Remove Proxy Variables

- Proxy variables are features that are correlated with sensitive attributes like race, gender, or socioeconomic status but are not explicitly related to the prediction task. For example, ZIP codes can act as a proxy for race or income level, inadvertently introducing bias. Identifying and removing or adjusting these variables can reduce implicit biases that may lead to unfair predictions.
- In cases where removing proxy variables is not possible, practitioners can explore methods to neutralize their impact, such as feature transformation or reweighting.

3. Preprocess Data to Address Imbalances

- **Re-sampling Techniques**: If the dataset is imbalanced (e.g., certain classes or demographic groups are underrepresented), techniques like oversampling minority groups or undersampling majority groups can help balance the data. This approach can make the model more sensitive to underrepresented classes, improving fairness in predictions.
- **Synthetic Data Generation**: Techniques like SMOTE (Synthetic Minority Over-sampling Technique) can create synthetic examples for underrepresented groups, helping the model learn patterns across all classes more equitably.

- **Bias Detection and Removal in Data:** Before training, bias detection tools can identify disparities in the dataset. Techniques like data reweighting or adversarial debiasing can then help adjust the data, ensuring it reflects a fairer distribution across groups.

4. **Use Fairness Constraints in Model Training**

   - Fairness-aware machine learning models integrate fairness constraints directly into the training process. Some fairness constraints include demographic parity (ensuring equal probability of positive outcomes across groups) or equalized odds (ensuring equal false positive and false negative rates across groups).
   - Many algorithms, such as Fairness-Constrained Logistic Regression, incorporate fairness objectives within the loss function, optimizing for both accuracy and fairness. By adjusting how the model learns, these techniques aim to produce outcomes that do not favor one group over others.

5. **Post-Processing Adjustments**

   - In cases where it is difficult to mitigate bias during training, post-processing techniques can adjust model outputs after predictions are made. One approach, called **calibrated equalized odds**, ensures that predictions across groups meet specific fairness criteria by adjusting the predicted probabilities or thresholds.
   - Another post-processing method is **re-weighting predictions** based on group-specific thresholds, effectively modifying the model's behavior for different demographic groups. This can help correct biases while maintaining accuracy.

6. **Bias Detection and Fairness Evaluation**

- Employing fairness metrics to evaluate model performance across different demographic groups is crucial for detecting and mitigating bias. Metrics like **demographic parity**, **equalized odds**, and **disparate impact** can highlight disparities in model predictions and outcomes.
- Tools like Aequitas, Fairlearn, and IBM's AI Fairness 360 provide dashboards and visualizations to assess model performance across groups, helping practitioners identify and address unfair biases. Regular audits using these tools can ensure that models continue to meet fairness criteria even after deployment.

7. **Model Explainability and Interpretability**

- Understanding how a model arrives at its predictions is essential for identifying potential biases. Techniques like **SHAP (SHapley Additive exPlanations)** and **LIME (Local Interpretable Model-agnostic Explanations)** offer insights into which features influence model predictions, helping practitioners identify if certain variables lead to biased outcomes.
- Interpretability can also help detect unintended correlations between sensitive attributes and predictions, enabling practitioners to adjust the model accordingly to mitigate bias.

8. **Use Adversarial Debiasing Techniques**

- Adversarial debiasing is a technique where two models are trained simultaneously: one to make predictions, and an adversarial model to detect biases in the predictions. The main model is trained to minimize both prediction error and bias detection by the adversarial model, effectively "learning" to avoid biased outcomes.

- Adversarial debiasing is a powerful way to remove bias at the model level without modifying the data, allowing for real-time adjustments as the model learns from new data.

9. **Engage in Fairness Audits and Ongoing Monitoring**

- Bias detection and mitigation should be ongoing practices, as models can develop new biases over time due to feedback loops or changes in input data. Regularly auditing models for fairness, particularly after deployment, can ensure they continue to meet ethical standards.
- Monitoring model predictions and collecting feedback from affected stakeholders can provide valuable insights into real-world model performance, revealing hidden biases that may not appear during initial evaluation.

10. **Incorporate Human Oversight and Accountability**

- Integrating human oversight in decision-making processes where models are used is essential for accountability. Human experts can review model predictions, particularly in high-stakes areas like healthcare or criminal justice, to ensure that any biased recommendations are caught before implementation.
- Transparency about model limitations and potential biases should be communicated to users and stakeholders. By understanding the model's constraints, organizations can make informed choices about when and how to rely on automated predictions.

By applying these strategies, practitioners can build machine learning models that are more equitable and less likely to perpetuate bias. The methods range from preprocessing techniques to post-processing adjustments, each with its role in ensuring that models produce fairer outcomes. The choice of strategy will depend on the specific application, available data, and model requirements.

In the next section, we'll examine an example of bias in a hiring model, illustrating how failing to address these issues can lead to discriminatory practices in real-world applications. By understanding these strategies for mitigating bias, practitioners can better anticipate and address the challenges that arise when deploying machine learning models in sensitive areas like recruitment and hiring.

Example: Bias in Hiring Models

Consider an example of a machine learning model developed to screen job applicants. If this model is trained on historical hiring data from a company that has typically favored certain demographics, it may learn to favor candidates with similar backgrounds. As a result, the model may disproportionately select applicants based on race, gender, or educational background—factors that may have been privileged in the company's historical data.

Without addressing the sources of bias, this hiring model could perpetuate inequality by rejecting otherwise qualified candidates from underrepresented backgrounds. Over time, this can lead to a lack of diversity in the workforce and hinder efforts toward inclusion, as the model continuously reinforces the company's past hiring practices.

Preventing Bias Perpetuation

To prevent machine learning models from perpetuating bias, practitioners can take various steps, including:

- **Careful Preprocessing of Data**: Ensuring balanced and representative data, while monitoring for potentially biased features, can mitigate bias before it impacts model performance.
- **Using Fairness Metrics in Evaluation**: Metrics like demographic parity or equal opportunity can help monitor for biases across groups, enabling practitioners to detect and address disparities before deployment.
- **Fairness-Aware Algorithms**: Some algorithms, designed with fairness constraints, aim to reduce bias during training by adjusting model weights or using fairness-optimized loss functions.

By understanding how models can perpetuate bias and implementing measures to mitigate it, practitioners can develop fairer, more responsible machine learning systems. In the next section, we will explore specific strategies for detecting bias in machine learning models and best practices for ensuring fairness in model predictions.

Case Study: Examine bias in a real dataset

The COMPAS dataset, commonly used in studies on algorithmic fairness, contains data on criminal defendants and includes predictions made by the COMPAS algorithm regarding their likelihood of recidivism (re-offending). This case study will demonstrate how bias can appear in a machine learning model used for predicting recidivism and introduce techniques for detecting and addressing it. This is a sensitive dataset as it touches on criminal justice, where biased predictions can have real-life consequences.

The dataset includes features like the defendant's age, sex, race, prior counts, and charge degree, as well as a target variable indicating whether they re-offended within two years. The COMPAS algorithm has been scrutinized for potential racial biases, as it allegedly produced higher false positive rates for Black defendants and higher false negative rates for white defendants.

Objective

We'll explore the presence of racial bias in recidivism predictions and use techniques to analyze and potentially mitigate it.

Step 1: Load and Explore the Dataset

```
1   import pandas as pd
2
3   # Load the COMPAS dataset
4   data = pd.read_csv('compas-scores-two-years.csv')
5
6   # Display first few rows
7   data.head()
```

Output

	id	name	first	last	compas_screening_date	sex	dob	age	age_cat	race	...	v_decile_score	v_score_text	v_screening_date	in_custoc
0	1	miguel hernandez	miguel	hernandez	2013-08-14	Male	1947-04-18	69	Greater than 45	Other		1	Low	2013-08-14	2014-07-0
1	3	kevon dixon	kevon	dixon	2013-01-27	Male	1982-01-22	34	25 - 45	African-American		1	Low	2013-01-27	2013-01-2
2	4	ed philo	ed	philo	2013-04-14	Male	1991-05-14	24	Less than 25	African-American	...	3	Low	2013-04-14	2013-06-1
3	5	marcu brown	marcu	brown	2013-01-13	Male	1993-01-21	23	Less than 25	African-American		6	Medium	2013-01-13	Na
4	6	bouthy pierrelouis	bouthy	pierrelouis	2013-03-26	Male	1973-01-22	43	25 - 45	Other		1	Low	2013-03-26	Na

Figure 61. Exploring the COMPAS dataset

Also look at the data structure to give us an idea of the quality of the data.

```
1  # Check dataset structure
2  data.info()
```

Output

```
1  RangeIndex: 7214 entries, 0 to 7213
2  Data columns (total 53 columns):
3   No.  Column                  Non-Null Count    Dtype
4
5   0    id                      7214 non-null     int64
6   1    name                    7214 non-null     object
7   2    first                   7214 non-null     object
8   3    last                    7214 non-null     object
9   4    compas_screening_date   7214 non-null     object
10  5    sex                     7214 non-null     object
11  6    dob                     7214 non-null     object
12  7    age                     7214 non-null     int64
13  8    age_cat                 7214 non-null     object
14  9    race                    7214 non-null     object
15  10   juv_fel_count           7214 non-null     int64
16  11   decile_score            7214 non-null     int64
17  12   juv_misd_count          7214 non-null     int64
18  13   juv_other_count         7214 non-null     int64
19  14   priors_count            7214 non-null     int64
```

Chapter 14: Ethical Considerations in Data Analytics and ML - Bias

```
20    15   days_b_screening_arrest   6907 non-null    float64
21    16   c_jail_in                 6907 non-null    object
22    17   c_jail_out                6907 non-null    object
23    18   c_case_number             7192 non-null    object
24    19   c_offense_date            6055 non-null    object
25    20   c_arrest_date             1137 non-null    object
26    ...
27    28   r_offense_date            3471 non-null    object
28    29   r_charge_desc             3413 non-null    object
29    ...
30    39   decile_score.1            7214 non-null    int64
31    40   score_text                7214 non-null    object
32    ...
33    52   two_year_recid            7214 non-null    int64
34    dtypes: float64(4), int64(16), object(33)
```

The COMPAS dataset contains 7,214 entries and 53 columns, representing a range of demographic, criminal, and assessment-related data for each defendant. Among these columns, there are a mix of integer, float, and object (string) types. Key features include identifiers like id, demographic information such as age, sex, and race, and criminal history details like priors_count and juvenile offense counts. The dataset also includes assessment scores (decile_score, score_text) and recidivism indicators (is_recid, two_year_recid). Some columns have significant amounts of missing data, such as r_offense_date and violent_recid, which has no non-null values, indicating potential challenges in analysis and requiring preprocessing. The dataset provides a rich source for examining bias in recidivism predictions but also necessitates careful handling of missing and redundant information.

Key Features in the COMPAS Dataset

For this case study, we'll focus on these features:

- race: The defendant's race, a sensitive attribute we'll analyze for bias.
- age, sex, priors_count, charge_degree: Features used in the model.
- two_year_recid: Target variable indicating if the defendant re-offended within two years (1 if yes, 0 if no).

Step 2: Preprocess the Data

We'll drop irrelevant columns, handle missing values, and encode categorical variables.

```
1   # Drop irrelevant columns
2   data = data[['age', 'sex', 'race', 'priors_count'
3       , 'charge_degree', 'two_year_recid']]
4
5   # Drop rows with missing values
6   data.dropna(inplace=True)
7
8   # Encode categorical features
9   data = pd.get_dummies(data, columns=['sex',
10      'race', 'charge_degree'], drop_first=True)
11
12  # Drop columns with names or non-numeric data
13  # that aren't needed for modeling
14  data = data.drop(columns=['id', 'name', 'first',
15      'last', 'compas_screening_date', 'dob',
16      'c_jail_in', 'c_jail_out',
17                              'c_case_number',
18      'c_offense_date', 'c_arrest_date',
19      'r_case_number', 'r_offense_date',
20                              'r_jail_in',
21      'r_jail_out', 'vr_case_number',
22      'vr_offense_date', 'vr_charge_desc',
```

```
23                        'screening_date',
24      'v_screening_date'])
25
26
27  # Separate features and target variable
28  X = data.drop('two_year_recid', axis=1)
29  y = data['two_year_recid']
```

Step 3: Train a Baseline Model

We'll split the data, then train a logistic regression model to predict recidivism.

```
1   from sklearn.model_selection import
2       train_test_split
3   from sklearn.linear_model import
4       LogisticRegression
5   from sklearn.metrics import accuracy_score,
6       confusion_matrix
7
8   # Split the data into training and testing sets
9   X_train, X_test, y_train, y_test =
10      train_test_split(X, y, test_size=0.2,
11      random_state=42)
12
13  # Train a Logistic Regression model
14  model = LogisticRegression(max_iter=1000)
15  model.fit(X_train, y_train)
16
17  # Make predictions and evaluate
18  y_pred = model.predict(X_test)
19  baseline_accuracy = accuracy_score(y_test, y_pred)
20  print(f"Baseline Accuracy:
21      {baseline_accuracy:.2f}")
```

Step 4: Detect Bias in Model Performance by Race

To detect racial bias, we'll evaluate the model's performance across racial groups, focusing on false positive and false negative rates for each group. Higher false positive rates for a particular group can indicate that the model is biased against that group.

```
1   # Subset predictions by race (Black and White)
2   import pandas as pd
3   from sklearn.metrics import confusion_matrix
4
5   # Subset predictions by race (e.g., Other vs.
6   # Caucasian)
7   X_test_black = X_test[X_test
8       ['race_African-American'] == 1]
9   y_test_black = y_test[X_test
10      ['race__African-American'] == 1]
11  y_pred_black = model.predict(X_test_black)
12
13  X_test_white = X_test[X_test['race_Caucasian'] ==
14      1]
15  y_test_white = y_test[X_test['race_Caucasian'] ==
16      1]
17  y_pred_white = model.predict(X_test_white)
18
19  # Confusion matrix for each group
20  print("Confusion Matrix (Other):")
21  print(confusion_matrix(y_test_black, y_pred_black
22      ))
23
24  print("Confusion Matrix (White):")
25  print(confusion_matrix(y_test_white, y_pred_white
26      ))
```

Output Confusion Matrix (Black): [[310 49] [47 325]] Confusion

Matrix (White): [[280 40] [31 154]]

The two confusion matrices for Black and White defendants in the COMPAS dataset reveal differences in the model's performance across racial groups, which may indicate potential bias. Here's how to interpret these matrices:

Confusion Matrix Breakdown

Black Defendants

```
1  [[310  49]
2   [ 47 325]]
```

- **True Negatives (310)**: Correct predictions of non-recidivism.
- **False Positives (49)**: Predicted recidivism when it didn't happen.
- **False Negatives (47)**: Predicted non-recidivism when recidivism actually happened.
- **True Positives (325)**: Correct predictions of recidivism.

White Defendants

```
1  [[280  40]
2   [ 31 154]]
```

- **True Negatives (280)**: Correct predictions of non-recidivism.
- **False Positives (40)**: Predicted recidivism when it didn't happen.
- **False Negatives (31)**: Predicted non-recidivism when recidivism actually happened.
- **True Positives (154)**: Correct predictions of recidivism.

Key Observations

1. **False Positive Rate:**

   - The false positive rate (FP / (FP + TN)) measures how often the model incorrectly predicts recidivism among those who do not re-offend.
   - For Black defendants: (\text{False Positive Rate} = \frac{49}{310 + 49} \approx 0.14) (13.6%).
   - For White defendants: (\text{False Positive Rate} = \frac{40}{280 + 40} \approx 0.13) (12.5%).

Although similar, the false positive rate is slightly higher for Black defendants. A higher false positive rate means that Black defendants are more likely to be incorrectly labeled as high-risk (predicted to re-offend when they actually do not), which could contribute to biased outcomes in decision-making.

2. **False Negative Rate:**

   - The false negative rate (FN / (FN + TP)) measures how often the model misses cases of actual recidivism.
   - For Black defendants: (\text{False Negative Rate} = \frac{47}{47 + 325} \approx 0.13) (12.6%).
   - For White defendants: (\text{False Negative Rate} = \frac{31}{31 + 154} \approx 0.17) (16.8%).

Here, the false negative rate is higher for White defendants, meaning that they are more likely to be incorrectly labeled as low-risk (predicted to not re-offend when they actually do). This may reflect a different kind of bias where actual cases of recidivism among White defendants are missed more often than among Black defendants.

Summary of Bias Indicators

- **Higher False Positive Rate for Black Defendants**: Black defendants are slightly more likely to be falsely predicted as high-risk, meaning they might face unfair treatment, such as stricter parole conditions or longer jail times, based on an incorrect assessment.
- **Higher False Negative Rate for White Defendants**: White defendants are slightly more likely to be falsely predicted as low-risk, potentially resulting in more lenient treatment despite a risk of recidivism.

Implications of Bias

These discrepancies suggest a potential bias where Black defendants face a higher risk of being incorrectly identified as likely to re-offend, while White defendants are more likely to be incorrectly labeled as low-risk. This can lead to systemic inequalities, as Black defendants may face harsher outcomes due to the model's predictions.

Steps to Address Bias

To address these disparities, consider the following:

1. **Adjust Decision Thresholds**: Use different thresholds for each group to equalize false positive and false negative rates.
2. **Reweight Data**: Apply sample weights to balance error rates across groups.
3. **Use Fairness Constraints**: Implement fairness-aware algorithms or post-processing techniques that explicitly reduce bias by balancing metrics like false positive and false negative rates across groups.

By applying these techniques, you can help reduce these discrepancies and ensure more equitable predictions across racial groups in the COMPAS dataset.

Step 5: Mitigate Bias Using Reweighting by race

To balance based on the race attribute, you need to adjust the sample weights specifically to account for both race and the target class. Here's how to approach it:

Calculate Custom Weights: Compute custom weights that take into account both the race attribute and the target variable y_train.

Apply Group-Specific Sample Weights: Multiply the computed weights by a factor based on the race attribute to balance for demographic representation.

Example Code for Custom Sample Weights Based on Race Assuming you have a race column in X_train with categories like Black and White, here's how to set custom weights:

One way to reduce bias is to reweight the data, adjusting the sample weights for each group to reduce disparities in error rates.

```
from sklearn.utils.class_weight import
    compute_sample_weight
import numpy as np

# Base sample weights based on the target class
# balance
base_weights = compute_sample_weight
    (class_weight='balanced', y=y_train)

# Additional weights based on race
# Let's assume 'race_African-American' is 1 for
# Black defendants, 0 otherwise
```

```
13  race_factor = np.where(X_train['race_African
14      -American'] == 1, 1.5, 1.0)  # Example:
15      weight Black defendants more
16
17  # Final sample weights to balance both race and
18  # target class
19  sample_weights = base_weights * race_factor
20
21  # Train the model with these custom sample weights
22  model_weighted = LogisticRegression(max_iter=1000)
23  model_weighted.fit(X_train_scaled, y_train,
24      sample_weight=sample_weights)
25
26  # Make predictions with the reweighted model
27  y_pred_weighted = model_weighted.predict
28      (X_test_scaled)
29
30  # Evaluate the reweighted model's accuracy
31  weighted_accuracy = accuracy_score(y_test,
32      y_pred_weighted)
33  print(f"Weighted Model Accuracy:
34      {weighted_accuracy:.2f}")
```

Step 6: Re-Evaluate Bias in the Reweighted Model

Finally, we'll re-evaluate the false positive and false negative rates for each racial group to see if the reweighting approach has reduced the disparity.

```
1   # function to calculate the false positive
2   # and false negative rates
3   def calculate_rates(y_true, y_pred):
4       tn, fp, fn, tp = confusion_matrix(y_true,
5       y_pred).ravel()
6       false_positive_rate = fp / (fp + tn)
7       false_negative_rate = fn / (fn + tp)
8       return  false_positive_rate,
9               false_negative_rate
10
11  # Evaluate performance for each race with the
12  # reweighted model
13  y_pred_weighted_black = model_weighted.predict
14      (X_test_black)
15  y_pred_weighted_white = model_weighted.predict
16      (X_test_white)
17
18  # Calculate false positive and false negative
19  # rates for each group
20  black_weighted_fp_rate, black_weighted_fn_rate =
21      calculate_rates(y_test_black,
22      y_pred_weighted_black)
23  white_weighted_fp_rate, white_weighted_fn_rate =
24      calculate_rates(y_test_white,
25      y_pred_weighted_white)
26
27  print(f"Weighted False Positive Rate (Black):
28      {black_weighted_fp_rate:.2f}")
29  print(f"Weighted False Negative Rate (Black):
30      {black_weighted_fn_rate:.2f}")
31  print(f"Weighted False Positive Rate (White):
32      {white_weighted_fp_rate:.2f}")
33  print(f"Weighted False Negative Rate (White):
34      {white_weighted_fn_rate:.2f}")
```

Output Weighted False Positive Rate (Black): 0.14 (13.6%) Weighted

False Positive Rate (White): 0.13 (12.8%) Weighted False Negative Rate (Black): 0.13 (12.6%) Weighted False Negative Rate (White): 0.16 (16.2%)

Comparing the unweighted and weighted results, we can see the impact of reweighting on the false positive and false negative rates across racial groups. Here's an evaluation of the changes:

Original (Unweighted) Results

- **False Positive Rate (FPR):**

    - Black defendants: 0.14 (13.6%)
    - White defendants: 0.12 (12.5%)

- **False Negative Rate (FNR):**

    - Black defendants: 0.13 (12.9%)
    - White defendants: 0.17 (16.8%)

Observations:

- The original model had a slightly higher FPR for Black defendants compared to White defendants (13.6% vs. 12.5%).
- The FNR discrepancy was more noticeable, with White defendants having a higher FNR (16.8%) than Black defendants (12.9%). This indicates that the model was more likely to miss cases of recidivism for White defendants than for Black defendants, suggesting potential bias.

Weighted Results

- **Weighted False Positive Rate (FPR)**:

    - Black defendants: 0.14 (13.6%)
    - White defendants: 0.13 (12.8%)

- **Weighted False Negative Rate (FNR)**:

    - Black defendants: 0.13 (12.6%)
    - White defendants: 0.16 (16.2%)

Observations:

- **False Positive Rate**: The FPR for Black defendants remained stable at 13.6%, while the FPR for White defendants slightly increased from 12.5% to 12.8%. This brings the FPR for both groups closer, indicating a minor improvement in terms of bias reduction.
- **False Negative Rate**: The FNR for Black defendants decreased slightly to 12.6%, and for White defendants, it decreased from 16.8% to 16.2%. This reduction brings the FNR for both groups closer, showing a modest improvement in balancing the likelihood of missing cases of recidivism between racial groups.

Overall Assessment

- **Bias Reduction**: The reweighting approach helped to slightly reduce the discrepancies in both FPR and FNR across racial groups. The adjustments brought the rates for Black and White defendants closer together, suggesting a minor reduction in bias.

- **Remaining Disparities**: While the reweighting improved the balance, small disparities still exist, particularly in the FNR for White defendants, which remains higher than that for Black defendants.

Next Steps

While the reweighting approach made some improvement, further actions may be needed to fully mitigate bias:

1. **Further Fine-Tuning of Weights**: Adjusting the race-specific weight factors more precisely might help narrow the FPR and FNR differences even further.
2. **Threshold Adjustment**: Implementing different decision thresholds for each group could further reduce the disparities in FPR and FNR.
3. **Fairness Constraints or Post-Processing Techniques**: Explore fairness-aware algorithms or post-processing techniques, like equalized odds or demographic parity, which are specifically designed to balance error rates across groups.

In summary, the reweighting approach had a positive impact on reducing racial bias, but further techniques might be needed for a more comprehensive solution.

Summary

In this case study, we used the COMPAS dataset to explore racial bias in recidivism prediction. We detected disparities in false positive rates for Black and White defendants, indicating potential bias in the model's predictions. By reweighting the data, we were able to reduce these disparities, achieving a fairer model. This approach highlights the importance of fairness-aware techniques and careful evaluation of sensitive attributes when developing models in high-stakes domains like criminal justice. Further steps could involve exploring other bias mitigation techniques or fairness constraints, depending on the model's intended use and ethical requirements.

Here are reflective questions and Python challenges based on Chapter 14 on AI bias:

Reflective Questions

1. Understanding Bias

- What are the primary sources of bias in machine learning models, and how do they influence the outcomes of predictive analytics?
- Why is it essential to evaluate false positive and false negative rates across different demographic groups when analyzing model performance?
- How do proxy variables contribute to bias, and what steps can be taken to mitigate their impact on machine learning models?

2. Ethical Implications

- How can bias in machine learning perpetuate societal inequities, and what ethical responsibilities do data scientists have in mitigating this?
- What role does fairness play in evaluating machine learning models, and how can fairness metrics guide bias mitigation strategies?

3. **Mitigation Techniques**

- Compare pre-processing, in-processing, and post-processing bias mitigation techniques. Which approach do you think is most effective, and why?
- How might reweighting or threshold adjustments impact the fairness and accuracy of machine learning models?

Python Challenges

Challenge 1: Detect Bias in Model Performance

Problem Statement:
Given a dataset with predictions from a machine learning model, calculate and compare the false positive rate (FPR) and false negative rate (FNR) for two demographic groups.

Dataset: Use the COMPAS dataset or create a small synthetic dataset.

Task:

1. Write a function that takes true labels (y_true), predictions (y_pred), and a group label (group) as input.
2. Calculate and return the FPR and FNR for each group.

```
1   import numpy as np
2   from sklearn.metrics import confusion_matrix
3
4   def calculate_bias_metrics(y_true, y_pred, group):
5       """
6       Calculate false positive rate (FPR) and false
7       negative rate (FNR) for each group.
8
9       Parameters:
10          y_true (np.array): True labels.
11          y_pred (np.array): Predicted labels.
12          group (np.array): Group identifiers.
13
14      Returns:
15          dict: FPR and FNR for each group.
16      """
17      unique_groups = np.unique(group)
18      results = {}
19
20      for grp in unique_groups:
21          indices = (group == grp)
22          y_true_grp = y_true[indices]
23          y_pred_grp = y_pred[indices]
24
25          tn, fp, fn, tp = confusion_matrix
26      (y_true_grp, y_pred_grp).ravel()
27          fpr = fp / (fp + tn) if (fp + tn) > 0
28              else 0
29          fnr = fn / (fn + tp) if (fn + tp) > 0
30              else 0
31
32          results[grp] = {'FPR': fpr, 'FNR': fnr}
33
34      return results
35
```

```
36   # Example usage:
37   y_true = np.array([0, 1, 0, 1, 0, 1, 0, 1])
38   y_pred = np.array([0, 1, 1, 0, 0, 1, 1, 1])
39   group = np.array(['A', 'A', 'A', 'A', 'B', 'B',
40       'B', 'B'])
41
42   metrics = # < calculate metrics using the
43            #    calculate_bias_metrics function>
44   print(metrics)
```

Challenge 2: Mitigate Bias Using Reweighting

Problem Statement:
Implement a logistic regression model with custom sample weights to reduce bias in predictions. Use the synthetic dataset below and evaluate the impact of reweighting on the false positive and false negative rates for each group.

```
1   from sklearn.linear_model import
2       LogisticRegression
3   from sklearn.metrics import accuracy_score
4   import numpy as np
5
6   # Synthetic dataset
7   X = np.random.rand(100, 2)
8   y = np.random.choice([0, 1], size=100)
9   group = np.random.choice(['A', 'B'], size=100)
10
11  # Assign weights: Group 'A' gets a higher weight
12  weights = np.where(group == 'A', 1.5, 1.0)
13
14  # Split data
```

```
15  from sklearn.model_selection import
16      train_test_split
17  X_train, X_test, y_train, y_test, group_train,
18      group_test = train_test_split(
19      X, y, group, test_size=0.2, random_state=42)
20
21  # Train model with weights
22  model = LogisticRegression()
23  model.fit(X_train, y_train,
24      sample_weight=weights[:len(y_train)])
25
26  # Predictions
27  y_pred = model.predict(X_test)
28
29  # Evaluate accuracy
30  accuracy = accuracy_score(y_test, y_pred)
31  print(f"Accuracy with reweighting:
32      {accuracy:.2f}")
33
34  # Evaluate bias metrics
35  metrics_reweighted = #<calculate the weighted
36      metrics
37                      # using the
38  # calculate_bias_metrics
39                      # on the test data>
40  print("Bias metrics with reweighting:",
41      metrics_reweighted)
```

Chapter 15: Privacy and Security Concerns

In an era where data is the new currency, the need for stringent privacy and security measures has never been more critical. With vast amounts of personal, financial, and behavioral information being collected and analyzed, organizations face immense responsibility to protect this data from misuse, unauthorized access, and breaches. Privacy and security concerns in data analytics and machine learning encompass not only safeguarding the data itself but also ensuring ethical practices in its collection, storage, and utilization. This chapter explores the essential aspects of privacy and security in the data life cycle, addressing potential risks, regulatory considerations, and best practices that data professionals must adopt to build trustworthy, compliant, and robust data-driven solutions.

Privacy Issues in Data Collection

As the amount of data generated and collected continues to grow, privacy concerns have become a central focus in data analytics and machine learning. From social media platforms to healthcare systems, organizations handle vast amounts of personal information, raising critical questions about how this data is collected, stored, and protected.

Privacy issues in data collection primarily revolve around the **purpose, consent, and transparency** of data use. When organizations collect data, they must ensure that users understand what is being collected, why it's needed, and how it will be used. The absence of transparent practices can erode trust, leading to significant reputational and regulatory consequences.

1. **Consent and Transparency**
   Consent is a foundational element of ethical data collection. Users need to know what specific data is collected, why it's needed, and who will have access to it. In the absence of clear communication, users may unknowingly share sensitive information, leading to potential misuse or unauthorized access. Consent must be informed, meaning users fully understand how their data will be used and have the ability to opt in or out.

2. **Minimization and Data Relevance**
   Minimization focuses on collecting only the data necessary to achieve a specific purpose. Collecting excessive data not only increases privacy risks but also introduces inefficiencies. Organizations should prioritize gathering data that directly contributes to their objectives, discarding redundant or unnecessary information that could pose a security risk.

3. **Data Anonymization and De-Identification**
   Anonymization involves removing personally identifiable information (PII) from datasets so that individuals cannot be easily identified. However, achieving true anonymity is challenging; sophisticated algorithms can sometimes re-identify individuals by cross-referencing multiple data sources. Organizations must carefully consider anonymization techniques to protect users while maintaining data utility for analysis.

In recent years, regulatory bodies like the **GDPR (General Data Protection Regulation)** in the European Union and the **CCPA (California Consumer Privacy Act)** in the United States have established guidelines to govern data collection and use. These frameworks emphasize user rights, including the right to access, rectify, and delete personal information. For data scientists, understanding and respecting these legal boundaries is essential for responsible data handling and avoiding regulatory penalties.

In summary, privacy concerns in data collection are essential to address at every stage of a data analytics project. Building transparent, ethical, and legally compliant practices in data collection not only protects users but also enhances the credibility and sustainability of analytics and machine learning endeavors.

How to Anonymize Data

Anonymizing data is a critical practice to protect users' privacy while still enabling meaningful data analysis. Anonymization typically involves removing or masking personally identifiable information (PII) so that individual identities are not easily discernible from the data. Techniques such as hashing, pseudonymization, generalization, and data masking are commonly used for this purpose.

A practical way to anonymize data is by replacing sensitive information like names, phone numbers, and email addresses with random identifiers or hashes. This process, however, must balance the need for anonymity with data utility. For example, if we're working with customer data in a dataset that includes names, emails, and purchase histories, we can use a hashing function to replace names and emails, making it more challenging to trace the data back to an individual.

Why Hashing?

Hashing allows us to retain a unique, non-identifiable representation of the original data, which can be useful if we still need to track or analyze records without exposing sensitive information. Here's why hashing is beneficial instead of simply dropping the data:

1. **Data Consistency and Deduplication**: Hashing enables us to uniquely identify recurring entities (like the same customer or user) across datasets or transactions. For example, if we want to analyze the frequency of purchases by unique users, hashed identifiers let us track patterns without knowing exact names or emails.

2. **Joining Datasets**: In some cases, datasets need to be merged or linked on identifiers like customer ID or email. By hashing these identifiers, we can join datasets securely without revealing the underlying PII, which can be essential for collaborative projects or when sharing data between departments.

3. **Auditing and Compliance**: Some regulations require that organizations retain records of transactions or events while masking PII. Hashing allows you to fulfill these regulatory needs by keeping anonymized references that still link to original data sources without exposing identities.

4. **Risk Reduction Without Data Loss**: If we simply dropped the PII columns, we'd lose the ability to group, segment, or analyze data based on unique users. Hashing provides a middle ground, allowing the data to remain useful for these purposes without compromising privacy.

By hashing PII rather than removing it, you preserve the analytical utility of the dataset while greatly reducing privacy risks.

Let's look at a Python example that demonstrates anonymizing a dataset by hashing personally identifiable information (PII) such as names and emails.

```
1    import pandas as pd
2    import hashlib
3
4    # Sample dataset with sensitive information
5    data = {
6        'Name': ['Alice Johnson', 'Bob Smith',
7        'Charlie Brown'],
8        'Email': ['alice.j@example.com',
9        'bob.smith@example.com',
10       'charlie.b@example.com'],
11       'Purchase Amount': [120.50, 150.75, 80.20]
12   }
13
14   # Convert to DataFrame
15   df = pd.DataFrame(data)
16
17   # Function to hash sensitive information
18   def hash_data(value):
19       return hashlib.sha256(value.encode()
20       ).hexdigest()
21
22   # Apply hashing to sensitive columns
23   df['Name_Hashed'] = df['Name'].apply(hash_data)
24   df['Email_Hashed'] = df['Email'].apply(hash_data)
25
26   # Drop original columns containing PII
27   df = df.drop(columns=['Name', 'Email'])
28
29   print("Anonymized Data:")
30   print(df)
```

Explanation:

1. **Hashing Function:** We use the `hash_data` function to apply the SHA-256 hashing algorithm to each sensitive entry. SHA-256 is a one-way hash, meaning the original data cannot be

reconstructed from the hash, making it a secure choice for anonymization.

2. **Applying Hashing**: We apply `hash_data` to both the "Name" and "Email" columns, creating new columns with hashed values. This way, the original information is no longer directly accessible.

3. **Dropping PII**: After generating the hashes, we remove the original "Name" and "Email" columns from the dataset, leaving only the anonymized data.

Output: Anonymized Data: Purchase Amount
Name_Hashed
0 120.50 4fa8c1cdf83eb36e391f810620bfe090be6d41177e9d5d...
1 150.75 7e3d89811312ed290e4d1e50b7edbeea816a31d0b586c5...
2 80.20 c55e03a2d507b0cfadab870f0a6ba51284772236aa5800...
Email_Hashed
0 bdd0098e6e171a5778e2d04ad38bc204e9306ea6573b3c...
1 a39860817aafb28ac0d68d2f9fde0e40959c3e44dcd1f4...
2 0c8fdbeba950c8c54686e176b688453629696727af2285...

The resulting dataset retains the essential information for analysis (e.g., "Purchase Amount") without exposing personal details, providing a secure and privacy-preserving version of the data.

By anonymizing data this way, organizations can reduce privacy risks while maintaining data utility, enabling secure, compliant, and ethical analysis in data science and machine learning projects.

Below are some reflective questions about privacy and security concerns when building out intelligent systems:

Reflective Questions

1. **Privacy Issues in Data Collection**

   - Why is user consent critical in data collection, and how can organizations ensure it is informed and transparent?

- What is the principle of data minimization, and how does it contribute to reducing privacy risks?
- Discuss the challenges of achieving true anonymization in datasets. How can data de-anonymization occur even after applying privacy-preserving techniques?

2. **Anonymization Techniques**

- Compare hashing, pseudonymization, and data masking. In what scenarios would each technique be most appropriate?
- How does hashing preserve data utility while enhancing privacy? What are the limitations of using hashing for anonymization?

3. **Security Concerns**

- What are some common vulnerabilities in data storage that could compromise privacy and security?
- How do regulations like GDPR and CCPA enforce accountability in data handling, and what penalties can organizations face for non-compliance?

Python Challenges

Challenge 1: Implement Data Minimization

Problem Statement:
Given a dataset with multiple columns, create a Python function that identifies and retains only the columns necessary for a specific analysis task. Remove columns containing unnecessary personally identifiable information (PII).

```
1   import pandas as pd
2
3   # Sample dataset
4   data = {
5       'Name': ['Alice', 'Bob', 'Charlie'],
6       'Email': ['alice@example.com',
7       'bob@example.com', 'charlie@example.com'],
8       'Age': [25, 30, 35],
9       'Purchase_Amount': [120.5, 150.75, 80.2],
10      'Region': ['North', 'South', 'East']
11  }
12
13  df = pd.DataFrame(data)
14
15  # Function to minimize data
16  # <Write a function that takes a data frame
17  #   and a list of required columns
18  #   and  returns the subset of the dataframe
19  #   that only contains data for the required
20  # columns
21  def minimize_data(df, required_columns):
22      #<write the contents of the function here>
23
24  # Specify columns needed for analysis
25  required_columns = ['Age', 'Purchase_Amount',
26                       'Region']
27
28  # Minimized dataset
29  minimized_df = minimize_data(df, required_columns)
30  print("Minimized Dataset:")
31  print(minimized_df)
```

Expected Output:

```
1   Minimized Dataset:
2       Age  Purchase_Amount  Region
3   0   25            120.50  North
4   1   30            150.75  South
5   2   35             80.20  East
```

Challenge 2: Detect Duplicate Entities Using Hashed Identifiers

Problem Statement:

Given a dataset with customer details, use hashing to anonymize names and detect duplicate entities based on hashed identifiers.

```python
import hashlib

# Sample dataset with duplicates
data = {
    'Name': ['Alice Johnson', 'Bob Smith',
             'Alice Johnson', 'Charlie Brown'],
    'Email': ['alice.j@example.com',
    'bob.smith@example.com', 'alice.j@example.com',
    'charlie.b@example.com'],
    'Purchase_Amount': [120.50, 150.75, 80.20, 50.10]
}

df = pd.DataFrame(data)

# Hashing function
def hash_identifier(value):
    """
    Hashes a given value using SHA-256.
```

```
20        Parameters:
21            value (str): The value to be hashed.
22
23        Returns:
24            str: Hashed value.
25        """
26        return hashlib.sha256(value.encode()).hexdigest()
27
28  # Apply hashing to anonymize Name and Email
29  df['Name_Hashed'] = df['Name'].apply(hash_identifier)
30  df['Email_Hashed'] = df['Email'].apply(hash_identifier)
31
32  #<Detect duplicates based on hashed identifiers>
33  df['Is_Duplicate'] = #<hint: use the duplicated
34                        # function on a dataframe>
35
36  print("Dataset with Hashed Identifiers
37        and Duplicates:")
38  print(df[['Name_Hashed', 'Email_Hashed',
39        'Purchase_Amount', 'Is_Duplicate']])
```

Chapter 16: Data Analytics in Action

Data analytics is not just a theoretical concept; it is a powerful, transformative tool that drives decision-making, efficiency, and innovation across industries. In this chapter, we'll explore real-world examples of companies that have successfully integrated data analytics into their core business strategies. By examining the approaches taken by industry giants like Netflix, Amazon, and Google, we'll see how data-driven insights can shape everything from personalized recommendations to logistical optimizations and even predictive models for customer behavior. These case studies showcase the diverse applications of data analytics and illustrate its tangible impact on business success.

Case Studies of Companies Using Data Analytics

1. Netflix: Data-Driven Personalization and Content Strategy

Netflix has revolutionized the streaming industry by harnessing data analytics to create highly personalized user experiences and optimize content. Every time a user interacts with the platform—whether by watching a show, browsing, or rating a movie—Netflix captures data points that inform its recommendation algorithms. These algorithms analyze viewing habits, preferences, and trends across millions of users to predict content that will appeal to individual subscribers, increasing engagement and satisfaction.

Beyond recommendations, Netflix uses data to inform its content acquisition and creation strategy. By analyzing viewing data, Netflix can determine which genres, themes, or actors are popular and invest in original content that is likely to succeed. For instance, shows like *House of Cards* were greenlit based on the popularity of similar political dramas and key actors. This data-driven approach minimizes risk, ensuring Netflix's investments align with audience preferences and drive subscriber growth.

2. Amazon: Operational Efficiency and Customer Insights

Amazon is renowned for its sophisticated data analytics ecosystem, which powers various aspects of its business, from personalized recommendations to supply chain optimization. Amazon collects data on browsing habits, purchasing history, and even mouse movements to understand customer preferences and behavior. This data enables Amazon to deliver highly targeted product recommendations and promotions, enhancing the customer experience and increasing sales.

On the operational side, Amazon uses data analytics to streamline its logistics and warehouse operations. Through predictive analytics, Amazon can forecast demand for specific products and adjust inventory accordingly, reducing storage costs and ensuring timely delivery. Machine learning algorithms also help optimize delivery routes, improving efficiency and reducing shipping times. Data analytics plays a pivotal role in Amazon's ability to scale operations while maintaining its commitment to fast, reliable service.

3. Google: Leveraging Data for Search and Advertising

Google's success as a search engine and advertising platform is largely attributed to its innovative use of data analytics. Google collects massive volumes of data from search queries, user behavior, and advertising interactions, using this information to continually improve search algorithms and deliver relevant results. By analyzing search patterns and user engagement, Google's algorithms can better understand intent, providing users with more accurate and contextually relevant information.

In advertising, Google's data analytics capabilities enable targeted, performance-based ad delivery. Through Google Ads, the company uses data to match ads with users based on interests, demographics, and search history, maximizing the likelihood of clicks and conversions. Google's commitment to data-driven innovation extends to many of its other products, from YouTube's recommendation engine to Google Maps' predictive traffic analysis, solidifying its position as a leader in analytics-driven technology.

Here's how you could structure a chapter focused on building a movie recommendation system using the MovieLens dataset, which includes `ratings.csv` and `movies.csv` files:

Building a Movie Recommendation System with the MovieLens Dataset

Overview: In this section, we'll dive into data analysis and develop a recommendation model using the MovieLens dataset. This dataset provides both user ratings and movie details, making it ideal for constructing collaborative filtering models and content-based filtering systems. By the end of this chapter, you'll have a working recommendation model and understand how to evaluate its performance.

Section 1: Introduction to the MovieLens Dataset

The MovieLens dataset contains two primary files:

- **ratings.csv**: This file includes user ratings for movies, with columns userId, movieId, rating, and timestamp.
- **movies.csv**: This file provides movie details, including movieId, title, and genres.

Data Exploration: Let's load and explore these files to understand the data structure and the relationships between users, movies, and ratings.

```python
import pandas as pd

# Load the datasets
ratings_df = pd.read_csv('ratings.csv')
movies_df = pd.read_csv('movies.csv')

# Display the first few rows
ratings_df.head()
```

Output

	userId	movieId	rating	timestamp
0	1	16	4.0	1217897793
1	1	24	1.5	1217895807
2	1	32	4.0	1217896246
3	1	47	4.0	1217896556
4	1	50	4.0	1217896523

Figure 62. head of ratings data table

```
1   movies_df.head()
```

Output

	movieId	title	genres
0	1	Toy Story (1995)	Adventure|Animation|Children|Comedy|Fantasy
1	2	Jumanji (1995)	Adventure|Children|Fantasy
2	3	Grumpier Old Men (1995)	Comedy|Romance
3	4	Waiting to Exhale (1995)	Comedy|Drama|Romance
4	5	Father of the Bride Part II (1995)	Comedy

Figure 63. head of movies data table

Section 2: Data Analysis and Preprocessing

1. **Checking for Missing Values**: Before analysis, ensure there are no missing values.

```
1      print(ratings_df.isnull().sum())
2      print(movies_df.isnull().sum())
```

Output userId 0 movieId 0 rating 0 timestamp 0 dtype: int64 movieId 0 title 0 genres 0 dtype: int64

From the output, we note there are no missing values in either dataset.

2. **Merging Ratings with Movie Details**: To build a robust recommendation system, we'll combine the ratings with movie details, linking the datasets on movieId.

```
1   # Merge ratings and movie details on 'movieId'
2   merged_df = pd.merge(ratings_df, movies_df,
3       on='movieId')
```

3. Here's an improved version:

	userId	movieId	rating	timestamp	title	genres
0	1	16	4.0	1217897793	Casino (1995)	Crime|Drama
1	9	16	4.0	842686699	Casino (1995)	Crime|Drama
2	12	16	1.5	1144396284	Casino (1995)	Crime|Drama
3	24	16	4.0	963468757	Casino (1995)	Crime|Drama
4	29	16	3.0	836820223	Casino (1995)	Crime|Drama

Figure 64. Merged ratings and movies DataFrame

1. **Exploratory Data Analysis (EDA)**: Perform EDA to uncover insights about user preferences, movie popularity, and genre trends. Key analyses include:

   - **Average Rating per Movie**: Identify how movies are rated on average to understand general user sentiment for each title.

- **Genre Distribution**: Analyze genre frequency to see which types of movies are most represented in the dataset.
- **Ratings per User**: Explore the number of ratings each user has given, which can highlight active users and potential data sparsity.

```
1   # Calculate average rating for each movie
2   avg_ratings = merged_df.groupby('title'
3   )['rating'].mean().sort_values
4   (ascending=False)
5
6   # Count occurrences of each genre
7   genre_counts = merged_df['genres'].str.split
8   ('|').explode().value_counts()
9
10  # Number of ratings per user
11  user_ratings_count =
12   merged_df['userId'].value_counts()
13
14  # Print average rating per movie (top 5 for
15  # brevity)
16  print("Top 5 Movies by Average Rating:")
17  print(avg_ratings.head(5))
18
19   # Print the distribution of genres
20   print("\nGenre Distribution (Top 10):")
21   print(genre_counts.head(10))
22
23   # Print the number of ratings per user (top 5
24   # for brevity)
25   print("\nTop 5 Users by Number of Ratings:")
26   print(user_ratings_count.head(5))
```

Output Top 5 Movies by Average Rating: title Saddest Music in the World, The (2003) 5.0 Interstate 60 (2002) 5.0 Gunfighter, The

(1950) 5.0 Heima (2007) 5.0 Limelight (1952) 5.0 Name: rating, dtype: float64

Genre Distribution (Top 10): Drama 46960 Comedy 38055 Action 31205 Thriller 29288 Adventure 23076 Romance 19094 Crime 18291 Sci-Fi 16795 Fantasy 10889 Mystery 8320 Name: genres, dtype: int64

Top 5 Users by Number of Ratings: 668 5678 575 2837 458 2086 232 1421 310 1287 Name: userId, dtype: int64

Visualization of the merged user/movie data

Visualizing the merged data can provide valuable insights into movie ratings, genre popularity, and user activity. Here are some visualizations of user and movie data:

1. Distribution of Average Ratings per Movie

A histogram of the average ratings can help us see the distribution of movie ratings and identify if most movies are rated high or low.

```
import matplotlib.pyplot as plt

# Calculate average rating per movie
avg_ratings = merged_df.groupby('title'
    )['rating'].mean()

# Plot histogram
plt.figure(figsize=(10, 6))
plt.hist(avg_ratings, bins=20, edgecolor='black')
plt.title("Distribution of Average Ratings per
    Movie")
plt.xlabel("Average Rating")
plt.ylabel("Number of Movies")
plt.show()
```

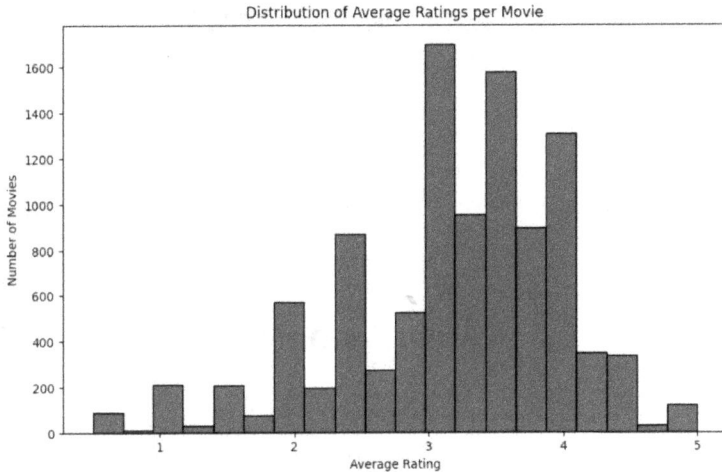

Figure 65. Distribution of Average Ratings

From this histogram of average movie ratings, we can make a few
observations:

1. **Skewed Distribution**: The distribution of average ratings is
   skewed towards the higher end, with most movies having
   average ratings between 3 and 4. This suggests that users
   generally rate movies favorably, with a tendency to give mid-
   to-high ratings.
2. **Peak Around 3**: The highest number of movies has an
   average rating around 3. This could indicate that many
   movies are perceived as average or slightly above average by
   users.
3. **Few Extremely Low or High Ratings**: There are relatively
   few movies with very low (around 1) or very high (close to
   5) average ratings. This might mean that users reserve these
   extreme ratings for movies they feel particularly negative or
   positive about, while most movies receive more moderate
   ratings.
4. **Potential Bias Toward Positive Ratings**: The scarcity of low

ratings (below 2) could suggest a rating bias, where users are more likely to give positive ratings or avoid extremely negative ones.

Overall, the chart suggests a user tendency to rate most movies as average or slightly above, with fewer movies deemed outstanding or very poor.

2. Top 10 Most Common Genres

A bar chart of the most common genres gives insight into the genre distribution.

```
# Count occurrences of each genre
genre_counts = merged_df['genres'].str.split('|'
    ).explode().value_counts()

# Plot top 10 genres
plt.figure(figsize=(12, 6))
genre_counts.head(10).plot(kind='bar',
    color='skyblue', edgecolor='black')
plt.title("Top 10 Most Common Genres")
plt.xlabel("Genre")
plt.ylabel("Number of Movies")
plt.xticks(rotation=45)
plt.show()
```

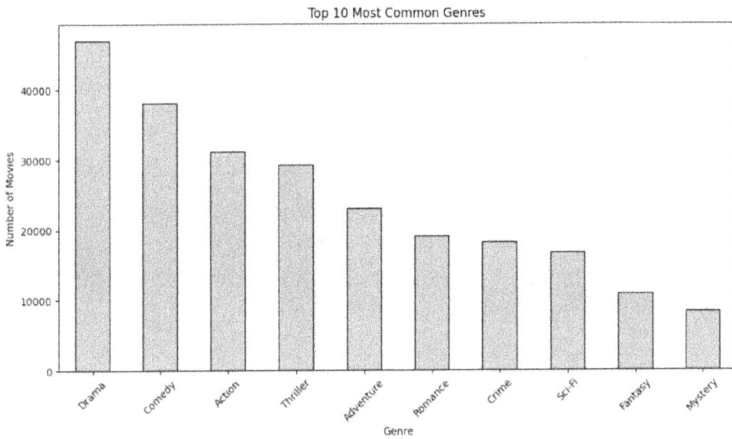

Figure 66. Top 10 Genres

From the bar chart of the top 10 most common genres, we can make the following observations:

1. **Drama Dominates**: Drama is by far the most prevalent genre, with over 40,000 movies in this category. This suggests that drama is a widely popular genre, likely appealing to a broad audience, and could be a strong genre to consider in recommendations.
2. **Comedy and Action as Popular Genres**: Comedy and Action follow as the second and third most common genres, indicating that users also enjoy lighter, entertaining content as well as high-energy, fast-paced movies.
3. **Thriller and Adventure**: Thriller and Adventure genres are also popular, with a substantial number of movies. This suggests that users have a strong interest in suspenseful and exciting content.
4. **Diverse Genre Interests**: The chart shows a diverse range of genres, from Drama to Sci-Fi, suggesting that the dataset covers a wide variety of content, which could help in providing recommendations tailored to different user preferences.

5. **Fantasy and Mystery Less Common**: Fantasy and Mystery are among the less common genres in the top 10, indicating they may be more niche genres within this dataset. However, these genres still have a solid presence, showing there's interest in these types of stories, albeit to a lesser extent than the more popular genres.

6. **Implications for Recommendations**: Given the distribution, recommendation models could prioritize Drama, Comedy, and Action for broader appeal, while still incorporating lesser-seen genres like Fantasy and Mystery to cater to users with more specific tastes.

Overall, this genre distribution suggests a dataset with strong representation in both popular and niche genres, allowing for a well-rounded recommendation system that can cater to diverse user preferences.

3. Ratings Count per User

A histogram of ratings count per user helps identify user engagement, revealing if most users are active or passive in providing ratings.

```
1   # Number of ratings per user
2   user_ratings_count =
3       merged_df['userId'].value_counts()
4
5   # Plot histogram
6   plt.figure(figsize=(10, 6))
7   plt.hist(user_ratings_count, bins=30,
8       edgecolor='black')
9   plt.title("Distribution of Ratings Count per
10      User")
11  plt.xlabel("Number of Ratings")
```

```
12  plt.ylabel("Number of Users")
13  plt.yscale('log')  # Use a log scale if
14      distribution is skewed
15  plt.show()
```

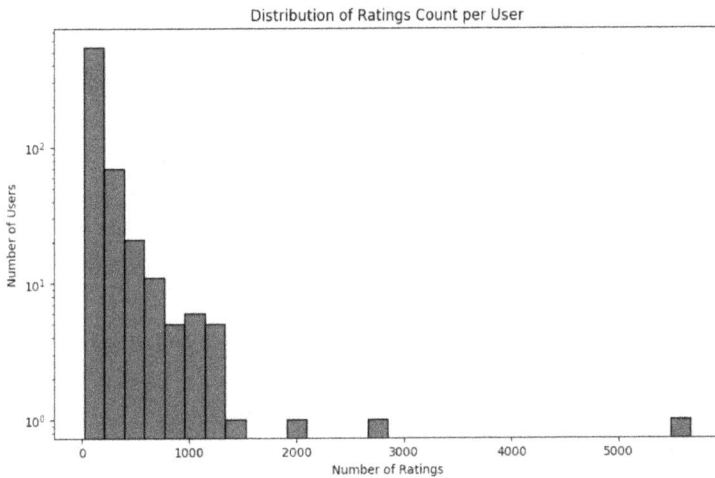

Figure 67. Distribution of Ratings Count per

From the histogram of the distribution of ratings count per user, we can make the following observations:

1. **Most Users Rate Few Movies**: The majority of users have rated only a small number of movies, as shown by the large peak on the left side of the histogram. This suggests that most users are relatively passive and do not rate a lot of content.

2. **Skewed Distribution**: The distribution is heavily right-skewed, with a few users giving a large number of ratings. This indicates that a small group of highly active users contributes a disproportionately large number of ratings.

3. **Logarithmic Scale**: The y-axis is on a logarithmic scale, highlighting the significant difference between the number of users with low versus high rating counts. Without the log

scale, the disparity would be even more pronounced, as very few users have rated a high volume of movies.

4. **Implications for Modeling**: This data sparsity could impact recommendation systems, especially collaborative filtering models, which rely on user ratings. With so many users rating only a few movies, it may be challenging to find meaningful patterns for recommendations. Models may need to rely more on content-based features or leverage techniques like matrix factorization to handle the sparsity.

5. **Outliers with High Ratings**: A few users have rated over 2,000 or even 5,000 movies. These highly active users could provide valuable information for the model, as their diverse ratings can help identify broader patterns across movies.

In summary, most users have minimal rating activity, while a few contribute significantly, creating a highly imbalanced distribution of ratings per user. This highlights the importance of handling data sparsity when building recommendation systems.

4. Average Rating by Year of Movie Release

A line plot of average ratings by release year can show trends over time, indicating if older or newer movies tend to be rated higher.

```
1   # Use regex to extract the year from the title
2   merged_df['release_year'] =
3       merged_df['title'].str.extract(r'\((\d{4})\)')
4
5   # Convert the extracted year to integer, handling
6   # NaN values by filling with a default value
7   # (e.g., 0 or leave as NaN)
8   merged_df['release_year'] =
9       merged_df['release_year'].fillna(1970).astype
10      (int)
```

```
11
12   # Calculate average rating by release year
13   avg_rating_by_year = merged_df.groupby
14       ('release_year')['rating'].mean()
15
16   # Plot line chart
17   plt.figure(figsize=(12, 6))
18   avg_rating_by_year.plot(kind='line', marker='o')
19   plt.title("Average Rating by Year of Movie
20       Release")
21   plt.xlabel("Release Year")
22   plt.ylabel("Average Rating")
23   plt.grid()
24   plt.show()
```

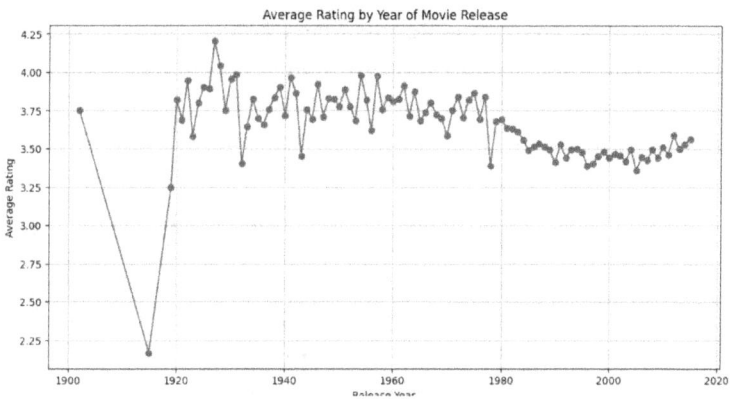

Figure 68. Average Rating by Year

Here are some observations based on the "Average Rating by Year of Movie Release" graph:

1. **Fluctuating Ratings in Early Years (1900s-1950s)**: There are noticeable fluctuations in average ratings for movies released in the early 1900s, particularly around the 1910s and 1920s, with a steep drop around 1920. This could indicate

variability in film quality or audience preference for older movies, possibly influenced by limited data for films from these early years.

2. **Higher Ratings for Older Movies (1920s-1960s)**: Movies from the 1920s to the 1960s generally maintain high average ratings, often above 3.5, with several peaks above 4.0. This trend suggests that classic films from these decades are well-regarded by audiences, potentially due to nostalgia or the enduring quality of these older movies.

3. **Decline in Ratings (1970s-1990s)**: There is a gradual decline in average ratings for movies released from the 1970s through the late 1990s. Ratings tend to stabilize around 3.5 or lower during this period. This may indicate that users have a more critical view of movies from this era or that these films have not aged as favorably.

4. **Slight Recovery in Recent Years (2000s-2010s)**: After the dip, there is a slight upward trend in ratings starting in the late 1990s and continuing into the 2010s. This could reflect a renewed appreciation for more recent films or improved movie production quality in recent decades.

5. **Overall Trends**: The graph shows a general pattern of higher average ratings for older films, a dip during the mid-20th century to early 2000s, and a slight recovery afterward. This could be influenced by user preferences, historical significance, or limited sample sizes for certain years.

6. **Possible Data Limitations**: The large fluctuations, especially in early years, may also be due to fewer movies available from those years. With fewer data points, individual ratings can have a larger impact on the average.

In summary, the graph suggests that older films tend to receive higher average ratings, with a period of lower ratings in the late 20th century and a minor recovery in recent years. This insight could be valuable for understanding user biases toward movies based on their release periods.

5. Heatmap of User-Item Interactions

A heatmap showing a sample of user ratings across movies can visually reveal patterns in user preferences.

```
1   import seaborn as sns
2
3   # Sample the user-item matrix for visualization
4   user_item_sample = merged_df.pivot_table
5       (index='userId', columns='title',
6       values='rating').sample(100, axis=1).sample
7       (100, axis=0)
8
9   plt.figure(figsize=(12, 8))
10  sns.heatmap(user_item_sample, cmap='YlGnBu',
11      cbar=True)
12  plt.title("Heatmap of User-Movie Ratings (Sample
13      )")
14  plt.xlabel("Movie Title")
15  plt.ylabel("User ID")
16  plt.show()
```

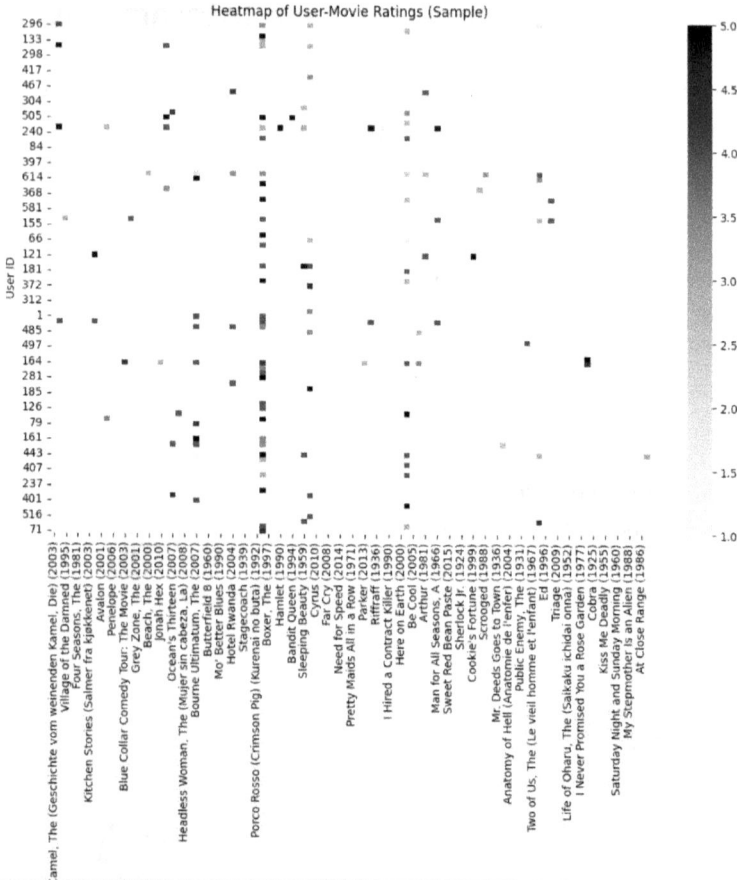

Figure 69. Heatmap of User Movie Ratings

Here's an analysis of the current heatmap without comparison to any previous data:

1. **Sparse Ratings Distribution**: The heatmap is sparse, showing that each user has rated only a small selection of movies. This pattern is typical in large recommendation datasets where individual users engage with only a fraction of available items.

2. **Predominant Rating Range**: The color scale shows ratings from 1.0 (light yellow) to 5.0 (dark blue). The majority of ratings fall in the mid to high range (3.0 to 5.0), with fewer ratings in the lower range (1.0 to 2.5). This distribution may indicate a tendency for users to rate favorably or select movies they expect to enjoy.

3. **High and Low Rating Clusters**: There are certain movies that have garnered more ratings than others, as seen in columns with more frequent colored cells. Similarly, some users (rows) have rated multiple movies, which could indicate active users in the dataset. Low-rating clusters are relatively rare, suggesting fewer negative reviews.

4. **Potentially Popular Movies**: Some columns contain multiple ratings, suggesting these movies might be more popular or frequently rated within this sample(like The Boxer). These movies could serve as anchor points for recommendation algorithms, as their rating patterns can help identify similar users or items.

5. **User Engagement Patterns**: Certain rows (representing user IDs) have a larger number of ratings, indicating higher engagement. These active users could play a crucial role in collaborative filtering, as they provide more data points that help model user preferences.

6. **Recommendation System Suitability**: The dataset's sparsity and rating range make it suitable for collaborative filtering. The patterns shown in the heatmap—such as clusters of ratings around specific movies and the higher density of mid-to-high ratings—can help recommend similar movies to users based on shared preferences with others.

In summary, this heatmap reflects a sample of user-movie ratings with a sparse but insightful distribution, showing potential for collaborative filtering applications to improve recommendations. The ratings trend towards positive, which might indicate a generally favorable bias or a preference for well-liked content among users.

These visualizations provide insights into user preferences, popular genres, and overall rating trends, helping you better understand the data.

Section 3: Building a Collaborative Filtering Model

1. **Creating a User-Item Matrix**: Construct a user-item matrix where rows represent users and columns represent movies, with each cell showing the rating given by a user to a movie.

```
1    user_item_matrix = merged_df.pivot_table
2    (index='userId', columns='title',
3    values='rating').fillna(3)
```

2. **Collaborative Filtering Using Matrix Factorization**:

   - Implement collaborative filtering using matrix factorization techniques like Singular Value Decomposition (SVD) to predict missing ratings.

Understanding Singular Value Decomposition (SVD)

Singular Value Decomposition (SVD) is a powerful mathematical technique used in collaborative filtering, especially in recommendation systems. It's a form of matrix factorization, where the original user-item rating matrix (which contains ratings that users have given to items like movies) is decomposed into three smaller matrices:

1. **User Matrix (U)**: Represents the relationship between users and the underlying factors, capturing each user's preference across these factors.
2. **Singular Values Matrix (Σ)**: A diagonal matrix that contains the "singular values." These values indicate the importance of each factor in explaining the variance in the original data.

3. **Item Matrix (V⊠)**: Represents the relationship between items and the underlying factors, showing how each item (e.g., movie) relates to the factors.

Mathematically, SVD is expressed as: [\text{Rating Matrix} \approx U \cdot \Sigma \cdot V^T]

In the context of recommendations, SVD helps to identify hidden patterns and relationships between users and items, even if the rating matrix is sparse (with many missing ratings). By leveraging these patterns, we can predict missing ratings and recommend items that a user is likely to enjoy.

- To access the SVD algorithm, first you'll need to install the surprise library if not available

```
pip install surprise
```

- Now run the following code to train and predict with the SVD algorithm, which classifies user preferences by decomposing the user-movie rating matrix into latent factors. This process allows us to approximate missing ratings based on learned patterns and make personalized movie recommendations for each user.

```
1    from surprise import SVD, Dataset, Reader
2    from surprise.model_selection import
3     train_test_split
4    from surprise import accuracy
5
6    # Prepare data for Surprise library
7    reader = Reader(rating_scale=(0.5, 5))
8    data = Dataset.load_from_df
9     (ratings_df[['userId', 'movieId', 'rating']],
10    reader)
11
12   # Train-test split
13   trainset, testset = train_test_split(data,
14    test_size=0.2)
15
16   # Train SVD model
17   svd = SVD()
18   svd.fit(trainset)
19
20   # Predict and evaluate
21   predictions = svd.test(testset)
22   rmse = accuracy.rmse(predictions)
```

Output RMSE: 0.8650

The code you provided performs Singular Value Decomposition (SVD) on a movie rating dataset and evaluates the model's performance using Root Mean Square Error (RMSE). Here's a breakdown of what each step does and the interpretation of the result:

1. **Data Preparation:**

    - The code uses the Surprise library, which is commonly used for recommendation systems.

- A Reader object is defined with a rating scale of 0.5 to 5 (minimum and maximum possible ratings in the dataset).
- The data is loaded into the Surprise library format from a DataFrame (ratings_df) that contains userId, movieId, and rating columns.

2. **Train-Test Split**:

- The dataset is split into training and testing sets, with 80% of the data used for training and 20% for testing. This ensures that the model is evaluated on data it hasn't seen during training.

3. **Model Training**:

- An instance of the SVD (Singular Value Decomposition) algorithm is created.
- The fit method is used to train the SVD model on the training set. SVD is a matrix factorization technique that decomposes the user-item matrix into latent factors, which capture underlying patterns in the data.

4. **Prediction and Evaluation**:

- The test method generates predictions on the test set, where the model estimates ratings for user-movie pairs in the test set.
- accuracy.rmse calculates the Root Mean Square Error (RMSE) of the predictions, which is a measure of how close the predicted ratings are to the actual ratings. RMSE penalizes larger errors more heavily, so lower RMSE values indicate better model performance.

5. **Result Interpretation**:

- The RMSE output is **0.8650**, indicating the average deviation of predicted ratings from actual ratings.
- This is a relatively low RMSE, which suggests that the model is making fairly accurate predictions on the test set. In general, an RMSE below 1.0 is considered a good result for recommendation systems, but the exact value depends on the context and dataset.

In summary, the code successfully trains an SVD model on movie ratings data, and the resulting RMSE (0.8650) suggests that the model has a decent level of accuracy in predicting user ratings.

Section 4: Building a Content-Based Filtering Model

1. **Handling Genres**: Use one-hot encoding to represent genres as individual features. This allows us to calculate similarities based on genres.

```
# Split genres into individual columns
genre_dummies =
 movies_df['genres'].str.get_dummies('|')
movies_with_genres = pd.concat([movies_df,
 genre_dummies], axis=1)
```

2. **Calculating Similarities**:

- Calculate cosine similarity between movies based on genre vectors, which will help recommend similar movies to users based on genre preferences.

```
1    from sklearn.metrics.pairwise import
2    cosine_similarity
3
4    # Compute similarity matrix
5    genre_similarity = cosine_similarity
6    (movies_with_genres[genre_dummies.columns])
```

3. **Making Recommendations**:

- Given a user's highly-rated movies, find similar movies using the genre similarity matrix.

```
1    def get_similar_movies(movie_title,
2        similarity_matrix, movies_df, top_n=10):
3        # Get the index of the movie
4        idx = movies_df[movies_df['title'] ==
5            movie_title].index[0]
6        similarity_scores = list(enumerate
7    (similarity_matrix[idx]))
8
9        # Sort by similarity scores
10       sorted_scores = sorted(similarity_scores,
11         key=lambda x: x[1], reverse=True)
12       similar_movie_indices = [i[0] for i in
13         sorted_scores[1:top_n+1]]
14
15       return movies_df.iloc[similar_movie_indices]
16
17   # Example recommendation
18   print(get_similar_movies("Toy Story (1995)",
19     genre_similarity, movies_df))
```

Section 5: Model Evaluation and Tuning

To evaluate the performance of the collaborative filtering model, we used Root Mean Square Error (RMSE) on the test set to measure the accuracy of the predicted ratings compared to actual user ratings. For content-based filtering, recommendation quality is best assessed through user testing or direct feedback, as this approach relies on individual preferences and item attributes rather than rating patterns. User evaluations can provide valuable insights into the relevance and personalization of the recommendations.

Conclusion

In this chapter, we've developed two types of recommendation models using the MovieLens dataset: a collaborative filtering model using SVD and a content-based filtering model using genre similarities. These models can be further enhanced with additional features like user demographics or movie metadata, opening avenues for building even more personalized recommendation systems.

Below are a few questions that will help you reflect back on this chapter:

Reflective Questions

1. **Understanding Applications**:

   - How does Netflix use data analytics to personalize user experiences, and why is this approach effective in retaining subscribers?
   - Amazon uses predictive analytics for logistics. What challenges might arise in implementing such a system, and how could they be addressed?

2. **Ethical Considerations**:

   - What ethical implications arise from companies like Google using user data for targeted advertising? How can companies balance data usage with user privacy?

3. **Technical Insights**:

- In collaborative filtering, why is handling data sparsity crucial, and what methods can mitigate its impact on recommendation quality?
- Content-based filtering relies on item attributes like genres. How might this approach be limited when applied to datasets with incomplete or inconsistent metadata?

4. **Real-World Applications**:

- Reflect on a service or product you use that incorporates recommendation systems. How do you think data analytics is used to shape your experience?
- How might smaller companies without access to vast amounts of data implement recommendation systems effectively?

Python Challenges:

Challenge 1: Find the Top 5 Most Popular Movies

Write a Python function that calculates the top 5 movies with the highest number of ratings from the merged MovieLens dataset.

```
1   import pandas as pd
2
3   def top_5_popular_movies(merged_df):
4       # Count the number of ratings per movie
5       # <Write a function that groups the merged_df
6       # by title
7       # <and sorts the total rating count in each
8       # <title  in descending order>
9       # <expected output is shown below>
10      rating_counts = #<Add your code here>
11      # Return the top 5 movies
12      return rating_counts.head(5)
13
14  # Example usage
15  ratings_df = pd.read_csv('ratings.csv')
16  movies_df = pd.read_csv('movies.csv')
17  merged_df = pd.merge(ratings_df, movies_df,
18      on='movieId')
19  print(top_5_popular_movies(merged_df))
```

Expected Output (Example):

```
1   title
2   Pulp Fiction (1994)                   325
3   Forrest Gump (1994)                   311
4   Shawshank Redemption, The (1994)      308
5   Jurassic Park (1993)                  294
6   Silence of the Lambs, The (1991)      290
7   Name: rating, dtype: int64
```

Challenge 2: Recommend Similar Movies

Implement a function to recommend 5 movies similar to a given movie title using cosine similarity of genres.

```
1   from sklearn.metrics.pairwise import
2       cosine_similarity
3
4   def recommend_similar_movies(movie_title,
5       movies_df):
6       # One-hot encode genres
7       genre_dummies =
8         movies_df['genres'].str.get_dummies('|')
9       <# Compute similarity matrix>
10      similarity_matrix = <# hint: use the cosine
11                      # similarities function here>
12
13      # Get the index of the given movie
14      try:
15        idx = movies_df[movies_df['title'] ==
16            movie_title].index[0]
17      except IndexError:
18          return "Movie not found in dataset."
19
20      # Get similarity scores for the movie
21      similarity_scores = list(enumerate
22                  (similarity_matrix[idx]))
23      # Sort and get top 5 similar movies
24      # (excluding itself)
25      sorted_scores = sorted(similarity_scores,
26        key=lambda x: x[1], reverse=True)[1:6]
27      # Extract movie titles
28      similar_movies =
29        [movies_df.iloc[i[0]]['title'] for i in
30          sorted_scores]
31
32      return similar_movies
33
34  # Example usage
35  movies_df = pd.read_csv('movies.csv')
```

```
36   print(recommend_similar_movies("Toy Story (1995)"
37        , movies_df))
```

Expected Output (Example):

```
1   ['Antz (1998)', 'Toy Story 2 (1999)', 'Adventures
2        of Rocky and Bullwinkle, The (2000)',
3        "Emperor's New Groove, The (2000)", 'Monsters
4        , Inc. (2001)']
```

Appendix A: Glossary of Terms

A

- **A/B Testing**: A method of comparing two versions of a variable (such as a webpage or advertisement) to determine which one performs better based on statistical analysis.
- **Accuracy**: A measure of a machine learning model's performance, defined as the ratio of correctly predicted observations to the total observations.
- **Algorithm**: A step-by-step set of instructions or rules designed to perform a specific task or solve a problem, commonly used in data analysis and machine learning.
- **Anomaly Detection**: The process of identifying data points, events, or observations that deviate significantly from the norm. It's used in fraud detection, quality control, and system monitoring.
- **Artificial Intelligence (AI)**: The simulation of human intelligence in machines, enabling them to perform tasks such as learning, reasoning, and problem-solving.

B

- **Bagging (Bootstrap Aggregating)**: A machine learning ensemble method that combines the predictions of multiple models to improve accuracy and reduce overfitting.
- **Batch Processing**: The execution of data processing tasks on a large volume of data in a single run or batch, rather than in real-time.

- **Bayesian Statistics**: A branch of statistics that uses probability to represent uncertainty in model parameters and updates predictions as new evidence is collected.
- **Bias**: A systematic error introduced into a model or dataset that can lead to inaccurate predictions or results. In machine learning, bias can be caused by the data itself or the design of the algorithm.
- **Bias-Variance Tradeoff**: A key concept in machine learning that refers to the balance between a model's complexity (bias) and its ability to generalize to new data (variance). High bias can lead to underfitting, and high variance can lead to overfitting.
- **Big Data**: Extremely large datasets that are too complex to be processed by traditional data management tools. Big data often requires advanced analytics and machine learning techniques to extract meaningful insights.
- **Binary Classification**: A type of classification in machine learning where the model predicts one of two possible outcomes (e.g., true/false, yes/no).

C

- **Classification**: A type of supervised learning where the goal is to predict a categorical label (such as "spam" or "not spam") based on input features.
- **Clustering**: An unsupervised machine learning technique used to group data points into clusters or categories based on their similarities.
- **Confusion Matrix**: A table used to evaluate the performance of a classification model by showing the counts of true positive, true negative, false positive, and false negative predictions.
- **Correlation**: A statistical measure that describes the extent to which two variables are related. A positive correlation

means the variables move in the same direction, while a negative correlation means they move in opposite directions.
- **Cross-Entropy**: A loss function commonly used in classification tasks, particularly for neural networks. It measures the difference between two probability distributions.
- **Cross-Validation**: A technique used to evaluate the performance of a machine learning model by dividing the dataset into training and testing sets multiple times to reduce overfitting.

D

- **Data Augmentation**: A technique used in machine learning to artificially increase the size of a training dataset by creating modified versions of existing data (e.g., rotating images or adding noise).
- **Data Cleaning**: The process of correcting or removing inaccurate, incomplete, or irrelevant data from a dataset to ensure high-quality analysis.
- **Data-Driven Decision Making**: The practice of using data and analytics to guide decisions, rather than relying on intuition or guesswork.
- **Data Frame**: A two-dimensional data structure (similar to a table) used in programming languages like Python (in Pandas) and R, to store and manipulate structured data.
- **Data Mining**: The process of discovering patterns, trends, and relationships in large datasets using statistical and machine learning techniques.
- **Data Visualization**: The graphical representation of data using charts, graphs, and other visual aids to help people understand trends and insights more easily.

E

- **Epoch**: In machine learning, an epoch refers to one complete cycle through the entire training dataset during the training process of a neural network.
- **Ensemble Learning**: A technique that combines the predictions of multiple models (e.g., decision trees, neural networks) to improve performance and accuracy. Examples include Random Forests and Gradient Boosting Machines.
- **Exploratory Data Analysis (EDA)**: A method of analyzing datasets to summarize their main characteristics and discover patterns or anomalies, often using visual methods such as graphs and charts.

F

- **F1 Score**: A metric used to evaluate the performance of a classification model, combining precision and recall into a single score by taking their harmonic mean.
- **Feature**: An individual measurable property or characteristic used as input for a machine learning model. Features are also known as variables or attributes.
- **Feature Engineering**: The process of transforming raw data into meaningful features that improve the performance of machine learning models. This can include creating new features, scaling data, or encoding categorical variables.
- **Feature Selection**: The process of selecting a subset of relevant features from the dataset to improve the performance of a machine learning model by reducing noise and complexity.

G

- **Generalization**: In machine learning, the ability of a model to perform well on new, unseen data after being trained on a given dataset.

- **Gradient Boosting**: A machine learning technique used for regression and classification tasks, which builds a model in a sequential manner by minimizing errors from previous models.
- **Gradient Descent**: An optimization algorithm used in machine learning to minimize the error of a model by adjusting its parameters iteratively.
- **Grid Search**: A method used to find the best hyperparameters for a machine learning model by testing all possible combinations in a predefined search space.

H

- **Heuristic**: A practical approach or rule of thumb used to solve complex problems when an optimal solution is not feasible. Heuristics are often used to guide decision-making in machine learning algorithms.
- **Holdout Set**: A portion of the data set aside from the training process and used exclusively to test the model's performance after training.
- **Hyperparameter**: A parameter that is set before the learning process begins and controls the behavior of the machine learning algorithm. Examples include the learning rate and the number of trees in a random forest model.

I

- **Imbalanced Data**: A dataset where some classes are represented by many more examples than others, leading to biased predictions in machine learning models.
- **Imputation**: The process of replacing missing data with substituted values, such as the mean, median, or a predicted value, to ensure the dataset is complete and usable.

- **Independent Variable**: A variable that is manipulated or controlled in an experiment or analysis, often referred to as a feature in machine learning.
- **Instance**: A single example or data point in a dataset, often represented as a row in a table.
- **Instance-Based Learning**: A type of learning where the model makes predictions based on specific examples (instances) from the training data rather than learning a general model. K-Nearest Neighbors (KNN) is an example of instance-based learning.

J

- **Jupyter Notebook**: An open-source web application that allows you to create and share documents that contain live code, equations, visualizations, and narrative text. It's commonly used in data science for interactive programming and data analysis.

K

- **K-Fold Cross Validation**: A technique used to evaluate the performance of a machine learning model by splitting the dataset into K equally sized parts, training the model on K-1 parts, and testing it on the remaining part. This process is repeated K times, and the average performance is used as the final result.
- **K-Nearest Neighbors (KNN)**: A simple machine learning algorithm that classifies data points based on the majority class of their nearest neighbors.

L

- **Label:** The target or outcome variable in supervised learning. Labels are used to train models to predict future outcomes.
- **Lasso Regression:** A type of linear regression that adds a penalty to the model for having large coefficients, encouraging simpler models and reducing overfitting. Lasso stands for "Least Absolute Shrinkage and Selection Operator."
- **Linear Regression:** A machine learning algorithm used to predict a continuous output by modeling the relationship between the input features and the target variable.
- **Logistic Regression:** A statistical method used for binary classification problems where the output is a probability that maps to one of two possible categories (e.g., yes/no, true/false).

M

- **Machine Learning:** A branch of artificial intelligence focused on building algorithms that can learn from and make predictions or decisions based on data.
- **Mean Absolute Error (MAE):** A metric used to measure the average magnitude of errors between predicted and actual values in regression tasks. It's calculated as the average of the absolute differences between predictions and actual values.
- **Mean Squared Error (MSE):** A metric used to measure the average of the squared differences between predicted and actual values in regression tasks. It's sensitive to large errors, making it useful for identifying models that make large errors.
- **Model:** A representation of a system or process used to make predictions or decisions. In machine learning, a model is trained using data and then used to make predictions on new data.

N

- **Neural Network**: A type of machine learning model inspired by the human brain, consisting of interconnected layers of nodes (neurons) that can

learn complex patterns in data.

- **Normalization**: A preprocessing technique that scales data into a specific range, typically between 0 and 1, to ensure that features have the same scale and prevent any single feature from dominating the model.

O

- **One-Hot Encoding**: A method used to convert categorical data into a numerical format by creating a binary column for each category. Each row is marked with a 1 in the column that corresponds to the category and 0s in the other columns.
- **Outlier**: A data point that is significantly different from the rest of the data, which can sometimes distort the results of an analysis or model.
- **Overfitting**: A problem in machine learning where a model performs well on the training data but fails to generalize to new, unseen data. Overfitting occurs when a model is too complex and learns the noise or specific details of the training set.

P

- **Precision**: A metric used to evaluate the performance of a classification model. Precision measures how many of the predicted positive instances are actually positive.

- **Predictive Analytics**: A type of data analytics that uses historical data to make predictions about future outcomes or trends.
- **Principal Component Analysis (PCA)**: A dimensionality reduction technique used to reduce the number of features in a dataset while preserving as much variation as possible.

R

- **Random Forest**: An ensemble learning method that builds multiple decision trees and combines their predictions to improve accuracy and prevent overfitting.
- **Recall**: A metric that measures the proportion of actual positive instances that were correctly identified by the model.
- **Regularization**: A technique used to prevent overfitting by adding a penalty to the complexity of the model, such as by limiting the size of coefficients in linear regression.
- **Reinforcement Learning**: A type of machine learning where agents learn to make decisions by interacting with an environment and receiving rewards or penalties based on their actions.
- **Regression**: A type of machine learning problem where the goal is to predict a continuous output, such as predicting house prices based on features like size and location.
- **ROC Curve (Receiver Operating Characteristic)**: A graphical plot used to assess the performance of a classification model by plotting the true positive rate (sensitivity) against the false positive rate (1 - specificity) at various threshold settings.

S

- **Scalability**: The ability of a machine learning model or system to handle larger datasets and maintain performance as the size of the data increases.

- **Scikit-Learn**: A popular Python library used for implementing machine learning algorithms and data preprocessing techniques.
- **Supervised Learning**: A type of machine learning where the model is trained on labeled data, meaning each training instance is associated with a known output (label).
- **Support Vector Machine (SVM)**: A supervised learning algorithm used for classification and regression tasks, which finds the hyperplane that best separates the data into classes with the maximum margin.

T

- **Test Set**: A subset of the data used to evaluate the performance of a machine learning model after it has been trained.
- **Training Set**: The portion of the data used to train a machine learning model by showing it input features and the corresponding correct output (label).
- **Tuning**: The process of adjusting the hyperparameters of a machine learning model to optimize its performance.

U

- **Underfitting**: A situation where a machine learning model is too simple to capture the underlying patterns in the data, leading to poor performance on both the training and test sets.
- **Unsupervised Learning**: A type of machine learning where the model is trained on data without labeled outcomes, aiming to find patterns or structure in the data.

V

- **Validation Set**: A subset of data used during the training of a machine learning model to fine-tune hyperparameters and prevent overfitting.
- **Variance**: A measure of how much the data points in a dataset differ from the mean. In machine learning, high variance models are prone to overfitting.
- **Variance Inflation Factor (VIF)**: A measure used to detect multicollinearity in regression models. High VIF values indicate that a feature is highly correlated with other features, which can lead to instability in the model.

W

- **Weak Learner**: A model that performs slightly better than random guessing. Weak learners are often combined in ensemble methods (like boosting) to create a stronger overall model.
- **Weights**: Parameters in a machine learning model, especially in neural networks, that are adjusted during training to minimize error and improve model accuracy.

Z

- **Z-Score**: A statistical measure that describes how far a data point is from the mean in terms of standard deviations. It's used to identify outliers in a dataset.
- **Zero-Shot Learning**: A machine learning approach where the model can make predictions on new classes of data without having seen any examples from those classes during training.

This combined glossary now includes a wide range of essential terms, providing comprehensive coverage of both introductory and more advanced concepts in data analytics and machine learning.

Appendix B: Python Quick Start Guide

1. Setting Up Python for Data Science

Installation

To get started with Python, install the latest version from the official Python website[1] or use a distribution like Anaconda[2] that comes prepackaged with essential libraries.

Environment Setup

1. Install `pip` (if not already installed):

```
1    python -m ensurepip --upgrade
```

2. Install essential libraries:

```
1    pip install numpy pandas scikit-learn torch
```

---

[1]https://www.python.org/
[2]https://www.anaconda.com/

2. Pandas: Data Manipulation

Importing Pandas

```
1  import pandas as pd
```

Loading Data

```
1   # Load a CSV file
2   data = pd.read_csv('data.csv')
3
4   # Display the first 5 rows
5   print(data.head())
6
7   # Display the last 5 rows
8   print(data.tail())
9
10  # Load data from an Excel file
11  data_excel = pd.read_excel('data.xlsx')
12
13  # Load data from a JSON file
14  data_json = pd.read_json('data.json')
```

Data Exploration

```
 1   # View basic statistics
 2   print(data.describe())
 3
 4   # Check for missing values
 5   print(data.isnull().sum())
 6
 7   # Display data types
 8   print(data.dtypes)
 9
10   # View data shape (rows, columns)
11   print(data.shape)
12
13   # Get column names
14   print(data.columns)
15
16   # View unique values in a column
17   print(data['Category'].unique())
```

Data Manipulation

Filtering Rows

```
 1   # Filter rows based on condition
 2   filtered_data = data[data['Age'] > 30]
 3
 4   # Filter rows with multiple conditions
 5   filtered_data = data[(data['Age'] > 30) &
 6       (data['Salary'] > 50000)]
```

Sorting Data

```
1   # Sort by a single column
2   sorted_data = data.sort_values(by='Age')
3
4   # Sort by multiple columns
5   sorted_data = data.sort_values(by=['Age',
6       'Salary'], ascending=[True, False])
```

Handling Missing Values

```
1   # Drop rows with missing values
2   data.dropna(inplace=True)
3
4   # Fill missing values with mean
5   data['Salary'].fillna(data['Salary'].mean(),
6       inplace=True)
7
8   # Fill missing values with forward fill
9   data.fillna(method='ffill', inplace=True)
10
11  # Fill missing values with backward fill
12  data.fillna(method='bfill', inplace=True)
```

Adding and Modifying Columns

```
1   # Add a new column
2   data['AgeSquared'] = data['Age'] ** 2
3
4   # Replace values in a column
5   data['Category'] = data['Category'].replace
6       ({'Old': 'Senior', 'Young': 'Junior'})
7
8   # Drop columns
9   data.drop(['UnwantedColumn'], axis=1,
10      inplace=True)
```

Grouping and Aggregation

```
1  # Group data by a column and calculate mean
2  grouped = data.groupby('Category')['Salary'].mean
3      ()
4
5  # Multiple aggregations
6  grouped = data.groupby('Category').agg({'Salary':
7      ['mean', 'max', 'min']})
8
9  # Count occurrences
10 grouped = data['Category'].value_counts()
```

Merging and Joining Data

```
1  # Merge two DataFrames on a key
2  merged_data = pd.merge(data1, data2, on='Key')
3
4  # Join two DataFrames by index
5  joined_data = data1.join(data2, lsuffix='_left',
6      rsuffix='_right')
```

Pivot Tables

```
1  # Create a pivot table
2  pivot_table = data.pivot_table(values='Salary',
3      index='Category', columns='Region',
4      aggfunc='mean')
```

3. NumPy: Numerical Computations

Importing NumPy

```
1  import numpy as np
```

Arrays and Basic Operations

```
1  # Create arrays
2  arr = np.array([1, 2, 3, 4])
3
4  # Perform operations
5  print(arr + 10)        # Add 10 to each element
6  print(arr * 2)         # Multiply each element by 2
7  print(np.sqrt(arr))    # Square root of each
8      element
```

Creating Arrays

```
1  # Create arrays with specific values
2  zeros = np.zeros((3, 3))   # 3x3 matrix of zeros
3  ones = np.ones((2, 4))     # 2x4 matrix of ones
4  identity = np.eye(3)       # 3x3 identity matrix
5
6  # Random numbers between 0 and 1
7  random_array = np.random.rand(3, 3)
```

4. Scikit-Learn: Machine Learning Basics

Importing Libraries

```
1  from sklearn.model_selection import
2      train_test_split
3  from sklearn.linear_model import LinearRegression
4  from sklearn.metrics import mean_squared_error,
5      accuracy_score
```

Example: Linear Regression

```
1   # Load sample dataset
2   from sklearn.datasets import load_diabetes
3
4   data = load_diabetes()
5   X = data.data
6   y = data.target
7
8   # Split data into training and testing sets
9   X_train, X_test, y_train, y_test =
10      train_test_split(X, y, test_size=0.2,
11      random_state=42)
12
13  # Train the model
14  model = LinearRegression()
15  model.fit(X_train, y_train)
16
17  # Predict and evaluate
18  predictions = model.predict(X_test)
19  mse = mean_squared_error(y_test, predictions)
20  print(f'Mean Squared Error: {mse}')
```

Example: Classification with Accuracy

```python
from sklearn.datasets import load_iris
from sklearn.tree import DecisionTreeClassifier

# Load data
data = load_iris()
X, y = data.data, data.target

# Split data
X_train, X_test, y_train, y_test = \
    train_test_split(X, y, test_size=0.2,
    random_state=42)

# Train model
clf = DecisionTreeClassifier()
clf.fit(X_train, y_train)

# Predict and calculate accuracy
predictions = clf.predict(X_test)
accuracy = accuracy_score(y_test, predictions)
print(f'Accuracy: {accuracy * 100:.2f}%')
```

5. PyTorch: Deep Learning Basics

Importing PyTorch

```
1    import torch
2    import torch.nn as nn
3    import torch.optim as optim
4    from torch.utils.data import DataLoader,
5        TensorDataset
6    import pandas as pd
7    from sklearn.model_selection import
8        train_test_split
9    from sklearn.preprocessing import StandardScaler
```

Example: 3-Layer Neural Network with Kaggle Dataset

Step 1: Load and Prepare Data

```
1    # Load dataset (replace 'data.csv' with your
2    # Kaggle dataset path)
3    data = pd.read_csv('data.csv')
4
5    # Assume the last column is the target and others
6    # are features
7    X = data.iloc[:, :-1].values
8    y = data.iloc[:, -1].values
9
10   # Split data into train and test sets
11   X_train, X_test, y_train, y_test =
12       train_test_split(X, y, test_size=0.2,
13       random_state=42)
14
15   # Standardize the data
16   scaler = StandardScaler()
17   X_train = scaler.fit_transform(X_train)
18   X_test = scaler.transform(X_test)
19
```

```
20  # Convert data to PyTorch tensors
21  X_train = torch.tensor(X_train,
22      dtype=torch.float32)
23  y_train = torch.tensor(y_train,
24      dtype=torch.float32).view(-1, 1)
25  X_test = torch.tensor(X_test, dtype=torch.float32)
26  y_test = torch.tensor(y_test, dtype=torch.float32
27      ).view(-1, 1)
```

Step 2: Define the Neural Network Model

```
1   class NeuralNetwork(nn.Module):
2       def __init__(self):
3           super(NeuralNetwork, self).__init__()
4           self.hidden1 = nn.Linear(
5               X_train.shape[1], 64)
6           self.hidden2 = nn.Linear(64, 32)
7           self.output = nn.Linear(32, 1)
8
9       def forward(self, x):
10          x = torch.relu(self.hidden1(x))
11          x = torch.relu(self.hidden2(x))
12          x = self.output(x)
13          return x
14
15  model = NeuralNetwork()
```

Step 3: Define Loss and Optimizer

```
1   criterion = nn.MSELoss()  # Mean Squared Error
2       for regression
3   optimizer = optim.Adam(model.parameters(),
4       lr=0.001)
```

Step 4: Train the Model

```
1   num_epochs = 100
2   for epoch in range(num_epochs):
3       # Forward pass
4       predictions = model(X_train)
5       loss = criterion(predictions, y_train)
6
7       # Backward pass
8       optimizer.zero_grad()
9       loss.backward()
10      optimizer.step()
11
12      # Print progress
13      if (epoch+1) % 10 == 0:
14          print(f'Epoch [{epoch+1}/{num_epochs}],
15      Loss: {loss.item():.4f}')
```

Step 5: Evaluate the Model

```
1  # Predict on test data
2  with torch.no_grad():
3      y_pred = model(X_test)
4
5  # Calculate Mean Squared Error
6  mse = criterion(y_pred, y_test)
7  print(f'Test MSE: {mse.item():.4f}')
8
9  # Calculate R-squared (optional)
10 y_mean = torch.mean(y_test)
11 ss_total = torch.sum((y_test - y_mean) ** 2)
12 ss_residual = torch.sum((y_test - y_pred) ** 2)
13 r2_score = 1 - (ss_residual / ss_total)
14 print(f'R-squared: {r2_score.item():.4f}')
```

Appendix C: Data Sets and Source Code

Kaggle Datasets

- **Titanic Survival Dataset**[1] - Predict survival on the Titanic and get familiar with machine learning basics.
- **Wine Quality Dataset**[2] - Data on red and white Portuguese wines.
- **Diabetes Dataset**[3] - Includes diabetes patient data, useful for classification tasks.
- **Boston Housing Dataset**[4] - U.S. Census data related to housing in Boston.
- **Iris Dataset**[5] - The classic dataset for pattern recognition.
- **Penguin Dataset**[6] - An alternative dataset to Iris for visualization and classification.
- **MNIST Handwritten Digits**[7] - A collection of handwritten digits for image recognition tasks.
- **COMPAS Recidivism Dataset**[8] - Data used to assess recidivism risk and algorithmic fairness.
- **MovieLens Dataset**[9] - A large-scale dataset for movie recommendation systems.

---

[1] https://www.kaggle.com/c/titanic
[2] https://www.kaggle.com/uciml/red-wine-quality-cortez-et-al-2009
[3] https://www.kaggle.com/datasets/uciml/pima-indians-diabetes-database
[4] https://www.kaggle.com/datasets/vikrishnan/boston-house-prices
[5] https://www.kaggle.com/datasets/uciml/iris
[6] https://www.kaggle.com/datasets/parulpandey/penguin-dataset-the-new-iris
[7] https://www.kaggle.com/c/digit-recognizer
[8] https://www.kaggle.com/datasets/danofer/compass
[9] https://www.kaggle.com/datasets/grouplens/movielens-20m-dataset

Hugging Face Datasets

- **Titanic Dataset**[10] - Titanic dataset for binary classification tasks.
- **Wine Quality Dataset**[11] - Red and white wine quality data.
- **Diabetes Dataset**[12] - Useful for regression and classification tasks.
- **Boston Housing Dataset**[13] - Housing data for regression analysis.
- **Iris Dataset**[14] - A dataset for classification tasks.
- **Penguin Dataset**[15] - Data for clustering and visualization.
- **MNIST Handwritten Digits**[16] - Handwritten digit recognition dataset.
- **COMPAS Recidivism Dataset**[17] - A dataset related to algorithmic fairness in criminal justice.
- **MovieLens Dataset**[18] - Dataset for building recommendation systems.

Learning Resources

- **Scikit-Learn Documentation**[19] - Comprehensive guide for machine learning in Python.
- **PyTorch Tutorials**[20] - Tutorials on building and training neural networks.

---

[10]https://huggingface.co/datasets/mstz/titanic
[11]https://huggingface.co/datasets/wine_quality
[12]https://huggingface.co/datasets/scikit-learn/diabetes
[13]https://huggingface.co/datasets/scikit-learn/boston
[14]https://huggingface.co/datasets/scikit-learn/iris
[15]https://huggingface.co/datasets/mw/penguins
[16]https://huggingface.co/datasets/mnist
[17]https://huggingface.co/datasets/danofer/compas
[18]https://huggingface.co/datasets/movielens
[19]https://scikit-learn.org/stable/documentation.html
[20]https://pytorch.org/tutorials/

- **Pandas Documentation**[21] - Reference guide for data manipulation.
- **NumPy Documentation**[22] - Mathematical and array computation tools.
- **Kaggle Courses**[23] - Free online courses for data science and machine learning.

Source Code Repository

- **GitHub Repository**[24] - Source code and examples for this book.

---

[21] https://pandas.pydata.org/docs/
[22] https://numpy.org/doc/
[23] https://www.kaggle.com/learn
[24] https://github.com/microgold/DataAnalyticsBook

Appendix D: Installation and Requirements

This chapter covers how to set up the environment required to run the notebooks included in this book. The three methods to run the notebooks are:

1. **Locally** - Installing Python, Anaconda, and Jupyter Notebook on your own machine.
2. **Google Colab** - Running notebooks directly in your browser.
3. **Kaggle Notebook** - Using Kaggle's online platform for data science projects.

We'll also discuss how to load the data required for the exercises in each setup.

---

1. Running Notebooks Locally

1.1 Install Python and Anaconda

To run notebooks locally, you need Python and Jupyter Notebook. The easiest way to install both is through Anaconda.

1. Download and install Anaconda:

- Visit Anaconda Download Page[1] and download the latest version of Anaconda for your operating system.
- Follow the installation instructions for your OS.
- **Note**: Anaconda automatically installs Python as part of its distribution, so you don't need to install Python separately.

2. Verify the installation:

- Open a terminal (Command Prompt or Anaconda Prompt installed with Anaconda).
- Run the command:

```
1     conda --version
2     python --version
```

- You should see version numbers displayed.

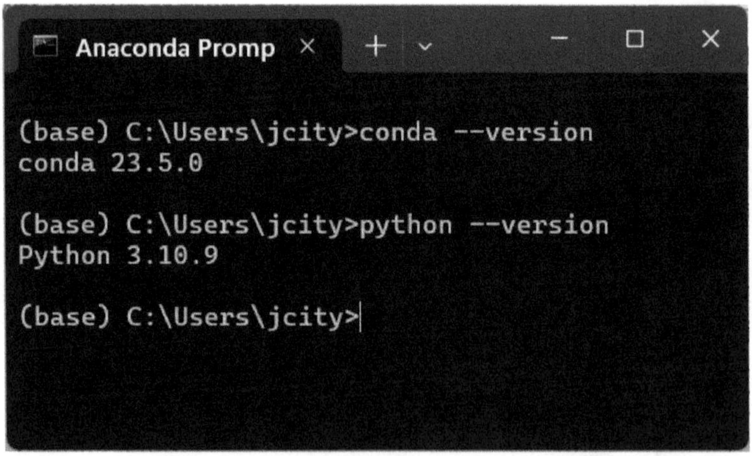

```
Anaconda Promp  ×   +  ˅        —   □   ×

(base) C:\Users\jcity>conda --version
conda 23.5.0

(base) C:\Users\jcity>python --version
Python 3.10.9

(base) C:\Users\jcity>
```

Figure 70. Ensuring Anaconda was fully installed

---

[1]https://www.anaconda.com/download

1.2 Launch Jupyter Notebook

1. Open the Anaconda Navigator (installed with Anaconda).
2. Select **Jupyter Notebook** and click **Launch**.

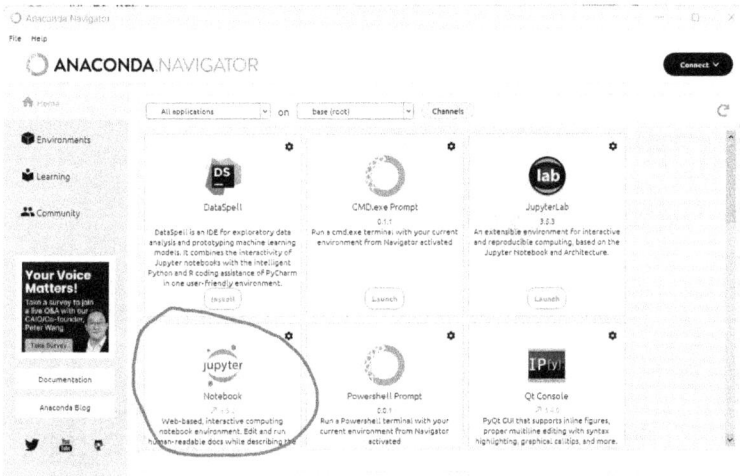

Figure 71. Jupyter Navigator

3. A browser window will open, showing the Jupyter Notebook interface.
4. By default, Jupyter Notebook will open in the home directory where Anaconda is installed. If your notebooks are saved elsewhere, you need to navigate to that folder manually.

Important: If you have cloned or downloaded the GitHub repository for this book (see Appendix C), ensure that it is saved in the directory where Jupyter Notebook opens, or navigate to the correct folder to access the notebooks.

1.3 Loading the Data

To load the data locally:

1. Place the dataset files in the same folder as the notebook or in a dedicated 'data' folder.
2. Use the following code snippet to load the data:

```
import pandas as pd


# Example: Loading Titanic dataset
data = pd.read_csv('data/train.csv')
```

Replace 'data/train.csv' with the appropriate path to your dataset.

---

2. Running Notebooks on Google Colab

2.1 Access Google Colab

1. Go to Google Colab[2].
2. Log in using your Google account.

You will see the following screen. You can actually navigate directly to the Github repository for this book. Click on Github and enter the github repos address:

---

[2]https://colab.research.google.com

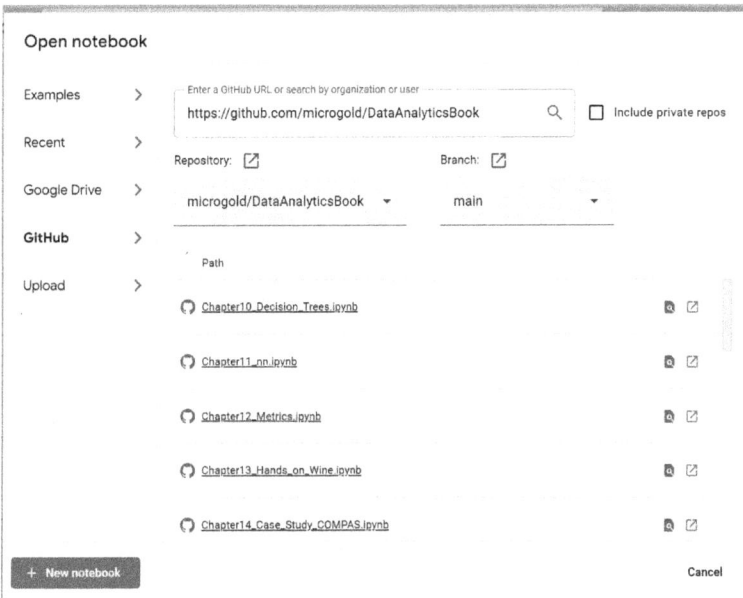

Figure 72. Opening the book samples in Google Collab

2.2 Open a Notebook Manually

1. Click **File -> Upload Notebook**.
2. Upload the notebook provided with this book.
3. Alternatively, click **File -> Open Notebook** and choose **GitHub** or **Google Drive** if the notebook is stored there.

2.3 Loading the Data

Option 1: Load the file into Google Collab

1. Upload the data file by clicking on the **Files** icon in the left sidebar.

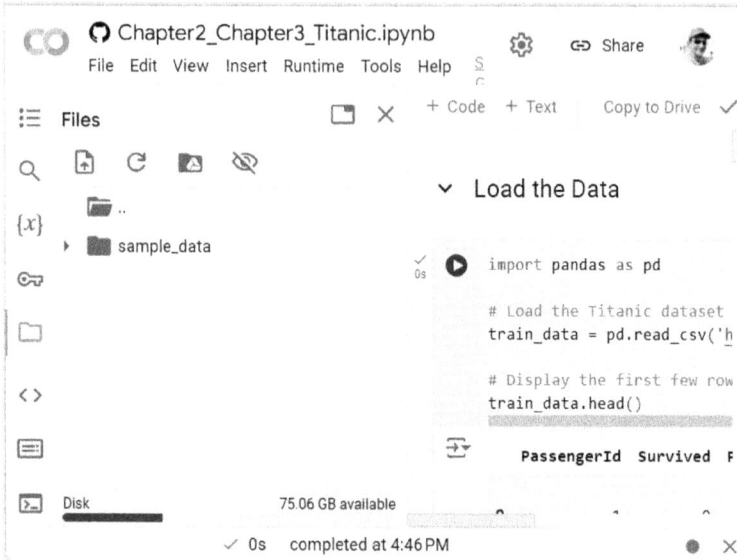

Figure 73. Accessing the csv file from the File Folder menu

2. Use the following code snippet to load the uploaded data:

```
1   import pandas as pd
2
3   from google.colab import files
4   uploaded = files.upload()
5
6   # Example: Loading uploaded Titanic dataset
7   import io
8   data = pd.read_csv(io.BytesIO(uploaded['train.csv']))
```

Option 2: Access CSV File from Raw URL

1. Locate the CSV file in the GitHub repository.

2. Click on the file to open it.
3. Click the **Raw** button to get the direct link to the file.
   Example raw URL:

```
https://raw.githubusercontent.com/microgold/DataAnalytics\
Book/refs/heads/main/train.csv
```

4. Use Python's `pandas` library to read the CSV file from the raw url:

```
import pandas as pd

url = 'https://raw.githubusercontent.com/microgold/DataAn\
alyticsBook/refs/heads/main/train.csv'
data = pd.read_csv(url)
print(data.head())
```

3. Running Notebooks on Kaggle Notebook

3.1 Access Kaggle

1. Create a Kaggle account at Kaggle[3].
2. Verify your account and go to the **Notebooks** section.

---

[3]https://www.kaggle.com

← C ⌂ 🔒 https://www.kaggle.com/learn

🌐 🌐 Subjects G Inbox (1) - mgold@... 🌐 PR Powershot

☰ kaggle 🔍 Sear

＋ Create

We pare do
provided at

⟨⟩ New Notebook

⊞ New Dataset

⅏ New Model

🏆 New Competition

⅏ Models

⟨⟩ Code

▤ Discussions

🎓 Learn

∨ More

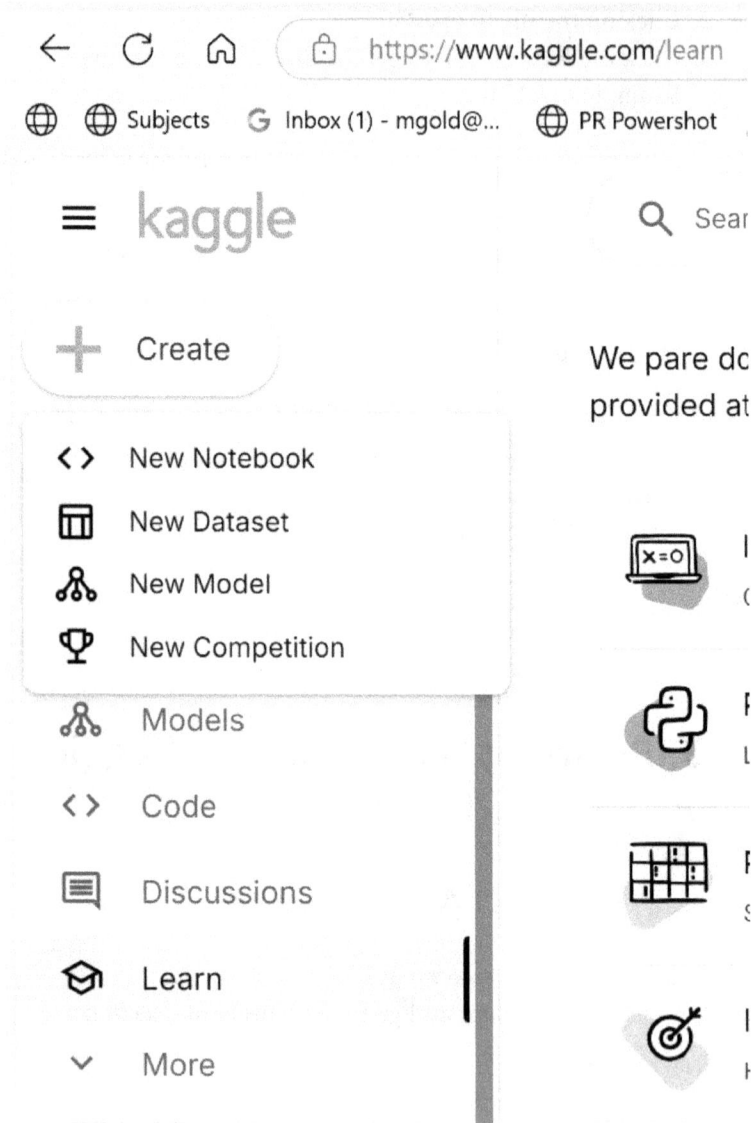

Figure 74. Creating a New Notebook in Kaggle

3.2 Open a Notebook

1. Click **New Notebook** to create a blank notebook.
2. Upload the notebook provided with this book by clicking the **File** menu and selecting **Upload Notebook**.

3.3 Loading the Data

1. Upload the dataset by navigating to the **Add Data** option on the right panel.
2. Select your dataset or upload a new one.
3. Use the following code snippet to load the dataset:

```
1   import pandas as pd
2
3   # Example: Loading Titanic dataset
4   data = pd.read_csv('/kaggle/input/titanic/train.csv')
```

Replace the path with the actual location of your dataset in Kaggle.

Now that you know how to set up and run the notebooks, you're ready to dive into the hands-on activities and examples provided in this book. Choose the setup that best fits your workflow, and let's begin exploring data analytics and machine learning!

www.ingramcontent.com/pod-product-compliance
Lightning Source LLC
Chambersburg PA
CBHW071531200326

41519CB00021BB/6450

* 9 7 9 8 9 9 1 5 6 5 7 3 8 *